ISRO's History

from Bicycle to Sun & Moon

Anand Shinde

Dedication

With profound gratitude, I dedicate this book to the orchestrator of my life's journey. Your blessings infuse every word, guiding the pages with the same grace that has illuminated my path.

!! Jai Mahakaal !!

About the Author

Anand Shinde is a dedicated cybersecurity expert with extensive experience in diverse cybersecurity technologies. His primary mission is to establish a safer Cyberworld through raising cybersecurity awareness initiatives. Anand's unwavering commitment is evident through his authorship of two pivotal books on cybersecurity, positioning him as a respected figure within the cybersecurity community. His impactful contributions have garnered numerous awards, highlightin g his exceptional influence in the field.

Beyond his literary achievements, Anand extends his passion for knowledge by guiding and mentoring students as a trusted advisor and career counsellor at institutions, including MID ADT University and various educational establishments. Prior to embracing his role as an author, Anand's professional journey spanned across some of the world's largest multinational companies, encompassing India, Poland, Ireland, and the United States of America. This extensive global experience equips him with a diverse perspective to address intricate cybersecurity challenges and offer practical solutions.

Anand's fascination with science and technology flourished from a young age, sparking an early interest in reading and a profound connection to Astronomy and space science. During his college years, he actively participated in scientific studies and events, including observing solar and lunar eclipses, meteor showers, and planetary occultations.

An observant enthusiast, Anand recognized the absence of comprehensive books encapsulating ISRO's rich history, evolutionary journey, and extraordinary accomplishments. This realization propelled him to embark on writing a book dedicated to ISRO's remarkable legacy.

For further engagement with Anand Shinde and to explore his extensive insights, visit his website at https://cyberauthor.tech/ His ardour for cyber security and space exploration encapsulates a journey of learning, mentoring, and authorship that continues to inspire and educate.

Table of content

Introduction

In the vast expanse of the cosmos, the Indian Space Research Organization (ISRO) has emerged as a shining star, captivating the minds and hearts of space enthusiasts worldwide. The world has been fascinated by the awe-inspiring achievements of ISRO, especially the resounding success of Chandrayaan-3, which has not only propelled ISRO to international acclaim but has also stirred a sense of immense pride among every Indian living in India and expats like me.

The triumph of Chandrayaan-3 has undoubtedly been a defining moment in India's space exploration journey. With each stride, ISRO has etched its name in the annals of history, significantly impacting the world's perception of India and its capabilities. This groundbreaking mission has transcended borders, instilling new dreams and igniting hope in the hearts of every Indian, both within the country and across its diaspora.

Although I am a Cyber Security Professional, I am also a devoted Astronomy enthusiast. I have noticed a lack of comprehensive books that genuinely encompass the vastness of ISRO's rich history, its evolutionary progress over time, and its exceptional missions and achievements. This realization has inspired me to embark on this literary voyage, where I aim to provide a holistic perspective on the evolution of ISRO from its humble beginnings to the groundbreaking success of Chandrayaan-3.

Join me in this celestial journey as we take a captivating look back in time, tracing ISRO's remarkable trajectory through the cosmos. Throughout the pages of this book, we shall explore the challenges and obstacles India faced in its pursuit of becoming a formidable space research explorer. From humble origins to becoming a pioneering space agency, ISRO's story is a testament to the indomitable human spirit and unwavering determination to explore the uncharted realms beyond our blue planet.

Our journey begins with ISRO's nascent inception, forged with a vision to harness space technology for India's development and progress. We shall trace the organization's humble beginnings as it pursued self-reliance and technological prowess in an era where space exploration seemed like an elusive dream. With each turning page, we will witness ISRO's steady ascent, propelled by innovative breakthroughs and leaps in space technology. From launching India's first satellite, Aryabhata, to mastering the art of satellite design and communication, ISRO's path has been illuminated by numerous milestones that have left an indelible mark on India's scientific landscape.

As we traverse further into the celestial depths of this book, we will encounter the awe-inspiring missions that have made ISRO a global space exploration powerhouse. Chandrayaan and Mangalyaan stand as shining examples of India's prowess in lunar and interplanetary missions, captivating the world with their success. The stories behind these missions testify to the indomitable spirit that defines ISRO's quest for scientific excellence. Moreover, we will explore ISRO's transformative impact on Indian society and economy. The applications of space technology have revolutionized communication, broadcasting, weather forecasting, agriculture, and disaster management, benefitting millions of lives across the nation. ISRO's contributions extend beyond Earth's atmosphere, opening new knowledge and space science frontiers.

While ISRO's achievements have garnered global attention and admiration, the journey has not been without its share of challenges. This book will shine a light on ISRO's hurdles and the determination that propelled it to surmount each obstacle, inspiring countless individuals to dream big and achieve the seemingly impossible.

Dear readers, in this literary endeavour, I wish to share the fascinating tales of the brilliant minds, the scientific ingenuity, and the unwavering pursuit of excellence that have made ISRO a source of immense pride for every Indian. Through these pages, I invite you to experience the celestial odyssey that has reshaped India's destiny among the stars and ignited the spirit of curiosity and exploration in the hearts of generations to come.

So, let us journey together, transcending time and space, as we explore the wonders of ISRO's space exploration saga - a journey that not only has made every Indian proud but also inspires humanity to look beyond the horizon and reach for the stars.

Astronomy in Ancient India

In the pursuit of understanding the historical work that transpired in ancient India on astronomy, one can trace back some of the earliest forms of astronomical knowledge to the period of the Indus Valley civilisation or even earlier. The ancient texts known as the Vedas hold cosmological concepts and insights into the movement of heavenly bodies and the passage of time. The Rig Veda describes time as a wheel with 12 parts and 360 spokes (representing days), along with a remainder of 5, hinting at the existence of a solar calendar.

During the early history of astronomy in India, a close association existed between this scientific pursuit and religious practices. The spatial and temporal requirements for the accurate performance of religious rituals necessitated astronomical observations. For example, texts like the Shulba Sutras, dedicated to altar construction, delved into advanced mathematics and fundamental astronomy.

Another significant text from this era is the Vedanga Jyotisha, which provided details about the Sun, Moon, nakshatras (lunar mansions), and the lunisolar calendar. The Vedanga Jyotisha also described rules for tracking the motions of celestial bodies, particularly the Sun and the Moon, to facilitate precise ritual timing. Within a yuga or "era," the text delineated parameters such as five solar years, 67 lunar sidereal cycles, 1,830 days, 1,835 sidereal days, and 62 synodic months.

In the 4th century BCE, Greek astronomical ideas began to permeate India due to the conquests of Alexander. This Indo-Greek influence continued manifesting in the astronomical tradition through texts like the Yavanajataka and Romaka Siddhanta. During this period, various siddhantas (treatises) were mentioned, with one such text being the Surya Siddhanta, although its precise content is not extant. The Surya Siddhanta, known today, dates back to the Gupta period and was received by the renowned mathematician and astronomer Aryabhata.

The classical era of Indian astronomy unfolded in the late Gupta era, from the 5th to 6th centuries CE. One of the notable works from this time is the Pañcasiddhāntikā authored by Varāhamihira around 505 CE. This treatise presented a method to determine the meridian direction using a gnomon based on three positions of the shadow. By the time of Aryabhata, planetary motion was treated as elliptical rather than circular, and various other astronomical topics, such as units of time and planetary motion models, were discussed.

The contributions of ancient Indian mathematicians and astronomers further enriched the field of astronomy. Baudhayan, for instance, calculated the value of Pi and provided equations resembling Pythagoras' theorem in his book Sulva Sutra.

Aryabhata, an independent thinker from the 5th century, calculated the value of Pi up to four decimals. His works spanned various disciplines, including mathematics, astronomy, physics, and more. In the Aryabhatiya, he discussed decimal systems, number theory, trigonometry, algebra, and astronomy. Notably, Aryabhata's invention of Zero revolutionised mathematics and allowed him to calculate the distance between the Earth and the Moon. He also proposed a heliocentric solar system model, arguing that solar and lunar eclipses resulted from this heliocentric system.

Brahmagupta, who lived around 500 years after Aryabhata, introduced new methods of multiplication in his book Brahmasphuta-siddhanta, which Arabs later translated. His writings included references to various measuring instruments used by astronomers to take readings.

In the 12th century, Bhaskaracharya emerged as a prominent mathematician born in Vijapur, Karnataka. His book Siddhanta Shiromani encompassed four sections: Lilavati (Arithmetic), Beej-Ganit (Algebra), Goladhyay (spheres), and Graha-Ganit (mathematics of planets).

Varahamihira, in his book Brihat-Samhita, wrote extensively about astronomy and the solar system. His other work, the Panchasiddhantika, described five astronomical systems.

The legacy of ancient Indian astronomers and mathematicians laid the foundation for the rich and profound astronomical tradition that later developed in India, contributing significantly to advancing human knowledge

4

about the cosmos. Their contributions continue to be celebrated and acknowledged in the study of astronomy and space exploration today.

Pre-Independence Work

The roots of modern space research in India can be traced back to a time before the country gained independence. In the 1920s, Mr. Sisir Kumar Mitra, a distinguished Physics professor at the University of Calcutta, embarked on a series of groundbreaking experiments aimed at studying the ionosphere. Utilizing ground-based radio technology in Kolkata, he conducted pioneering research on the sounding of the ionosphere.

Image source © insa.nic.in

Recognizing the immense potential of space research, Mr Sisir Kumar Mitra took a momentous step in 1949 by establishing a new department at the University of Calcutta. This department evolved into the prestigious Institute of Radio Physics and Electronics, with a core group of space physicists at its helm. Through their relentless efforts, the Institute paved the way for cutting-

edge research in the field of space science. Mr. Sisir Kumar Mitra's legacy continues to thrive through the S. K. Mitra Center for Research in Space Environment, which proudly bears his name at the University of Calcutta. This centre serves as a living tribute to his visionary contributions and serves as a hub for contemporary research in space environment studies. Mr Sisir Kumar Mitra's pioneering experiments and establishment of the Institute of Radio Physics and Electronics laid the foundation for the nation's later advancements in space exploration. His foresight and dedication to space physics contributed significantly to the development of a strong scientific ecosystem in India, fostering further exploration and research in space-related fields. The study of the ionosphere, initiated by Mr. Mitra, holds immense importance, as the ionosphere plays a crucial role in communication, satellite navigation, and space weather.

Meghnad Saha, a celebrated Indian astrophysicist, left an indelible mark on India's scientific landscape, playing a pivotal role in laying the groundwork for the nation's future space research endeavours. His groundbreaking contributions to astrophysics and profound impact on India's scientific community paved the way for the development of space research within the country. Meghnad Saha's most notable achievement was the formulation of the "Saha ionization equation," also known as the "Saha-Sommerfeld equation." This equation proved to be a game-changer, enabling scientists to comprehend the ionization state of matter under various conditions, particularly in stellar atmospheres. Its profound implications extended to advancing our understanding of stellar spectra, temperature, and composition of stars, revolutionizing the field of astrophysics.

Saha's pioneering work in astrophysics laid a strong foundation for the discipline's growth in India. His active involvement in esteemed academic and scientific institutions, such as the University of Calcutta and the Indian Association for the Cultivation of Science (IACS), where he served as director, further enriched the scientific ecosystem in the country. These institutions would later become instrumental centres of excellence, nurturing scientific research and education, including in the fields of space science and technology.

Saha's leadership and influence extended beyond the laboratory walls. He was the driving force behind the establishment of significant scientific societies, such as the National Academy of Science in 1930 and the Indian Physical Society in 1934, both serving as platforms for scientific collaboration and advancement. His vision also played a crucial role in establishing the prestigious Indian Institute of Science in 1935, which continues to stand as a symbol of India's commitment to scientific excellence.

Saha's profound impact on the scientific community and his unwavering dedication to the field were further evident during his tenure as the Director of the Indian Association for the Cultivation of Science from 1953 to 1956. His leadership brought forth new opportunities for scientific exploration and collaboration, leaving a lasting legacy within the organization. To honour his monumental contributions to science, the Saha Institute of Nuclear Physics, founded in Kolkata in 1943, bears his name, a fitting tribute to his extraordinary legacy.

Post-Independence Space Research

Homi Jehangir Bhabha, widely known as the "Father of Indian Nuclear Science" and one of the most prominent scientists in India's history, made significant contributions to nuclear research and laid the early foundations for space research in the country. Born on October 30, 1909, in Mumbai, Bhabha's brilliance and passion for scientific exploration would shape India's space and nuclear science trajectory.

After India's independence in 1947, Bhabha's vision for scientific progress was rooted in the belief that India should pursue nuclear and space research advancements, recognizing the transformative impact they could have on the nation's growth and development. One of Bhabha's earliest contributions to Indian space research was the establishment of the Tata Institute of Fundamental Research (TIFR) in 1945, even before India gained independence. TIFR became a hub for groundbreaking research in various fields, including nuclear physics, cosmic rays, and astrophysics. Bhabha's leadership and vision paved the way for TIFR to emerge as a premier research institute, nurturing scientific talent and laying the groundwork for space-related studies. In 1948, Bhabha founded the Cosmic Ray Research Unit at the Indian Institute of Science (IISc) in Bangalore, which played a crucial role in advancing cosmic ray research in India. This research unit provided valuable insights into high-energy cosmic ray particles and their interactions with Earth's atmosphere, contributing to the understanding of space-related phenomena. Bhabha's pioneering efforts in space research gained further momentum with the establishment of the Atomic Energy Commission (AEC) of India in 1948. He was appointed as its first chairman, providing him with a platform to spearhead nuclear research and space exploration initiatives.

Bhabha's vision for space research culminated in the establishment of the Indian National Committee for Space Research (INCOSPAR) in 1962. This committee was the precursor to the Indian Space Research Organisation (ISRO), officially formed in 1969. Under Bhabha's guidance and with support from Prime Minister Jawaharlal Nehru, INCOSPAR set the stage for India's foray into space exploration.

Homi Jehangir Bhabha's profound contributions to Indian space research, coupled with his visionary leadership, continue to inspire generations of scientists and space enthusiasts. His foresight in establishing institutions like TIFR, INCOSPAR, and the AEC set the stage for India's burgeoning space program.

D r. Vikram Sarabhai, a visionary scientist and institution builder, is often hailed as the Father of the Indian space program. His remarkable journey began when he returned to an independent India in 1947 after pursuing his studies at Cambridge. Fuelled by a deep sense of national pride and a passion for scientific progress, Sarabhai set out to establish institutions that would shape the scientific landscape of the nation.

On November 11, 1947, at the young age of 28, Vikram Sarabhai laid the foundation of the Physical Research Laboratory (PRL) in Ahmedabad. PRL was the stepping stone of his grand vision to create and nurture institutions that would drive scientific research and exploration in India. Initially, the focus at PRL was on cosmic rays and studying the properties of the upper atmosphere, but with grants from the Atomic Energy Commission, the research areas soon expanded to encompass theoretical physics and radio physics. Over the years, Sarabhai's leadership at PRL flourished, and his reputation as a creator and cultivator of institutions grew. His endeavours were not limited to space research; he played an important role in the establishment of numerous institutions across diverse fields. His foresight and dedication laid the groundwork for India's scientific progress in various domains.

Dr. Vikram Sarabhai emphasized the importance of a space program in his quote:

"There are some who question the relevance of space activities in a developing nation. To us, there is no ambiguity of purpose. We do not have the fantasy of competing with economically advanced nations in the exploration of the moon or the planets or manned space flight.

But we are convinced that if we are to play a meaningful role nationally, and in the community of nations, we must be second to none in the application of advanced technologies to the real problems of man and society."

In 1962, Sarabhai's persuasive prowess bore fruit as the Indian National Committee for Space Research (INCOSPAR) came into existence, spearheaded by Prime Minister Jawaharlal Nehru, who was swayed by Sarabhai's compelling vision for a space program in India. This laid the foundation for what would later become the Indian Space Research Organisation (ISRO). Dr. Sarabhai's dream of exploring the cosmos from Indian soil was taking shape. As a true champion of science education and outreach, Sarabhai founded the Community Science Centre in Ahmedabad in 1966. Today, this institution stands as the Vikram Sarabhai Community Science Centre, carrying forward his legacy of nurturing young minds and instilling a passion for science.

A pivotal moment in India's space journey came in 1975 when Sarabhai's vision to build and launch an Indian satellite materialized. Aryabhata, the first Indian satellite, was successfully placed into earth orbit in 1975 from a Russian cosmodrome. This historic achievement marked a significant milestone in India's space exploration endeavours and symbolized the nation's growing capabilities in space technology.

Dr Homi Jehangir Bhabha, renowned as the Father of India's nuclear science program, joined hands with Sarabhai in the pursuit of India's space ambitions. Together, they established India's first rocket launching station at Thumba, near Thiruvananthapuram, taking advantage of its strategic location near the equator. On November 21, 1963, all the efforts culminated in a momentous event - the inaugural flight that marked the beginning of India's modern rocketry program. The twilight sky over the fishing village of Thumba near Thiruvananthapuram witnessed a breathtaking sight as the orange trail illuminated the horizon. The excitement was palpable not only in Kerala but also in the neighbouring districts of Tamil Nadu. Such was the spectacle that even the proceedings of the Kerala Legislative Assembly were momentarily adjourned, allowing the members to witness the glorious display left behind in the western sky. The awe-inspiring sight was a result of the launch of the Nike Apache rocket, procured from the United States, and it captured the imagination of all who beheld it.

The Nike Apache rocket, a two-stage marvel, weighed 715 kg and operated on powerful solid propellants. As it soared to an altitude of 208 km, it released sodium vapour that painted the sky in a mesmerizing glow. This historic event marked the inception of India's journey into modern rocketry and laid the seeds for the nation's ambitious space exploration endeavours.

In 1966, Sarabhai's engagement with NASA resulted in the Satellite Instructional Television Experiment (SITE), which aimed to use satellites for instructional television. This groundbreaking project, launched between July 1975 and July 1976, showcased Sarabhai's commitment to leveraging advanced technologies for societal benefit. Vikram Sarabhai's contributions extended beyond India's borders, earning him international recognition. In 1974, the International Astronomical Union in Sydney honoured his legacy by designating the Moon Crater Bessel in the Sea of Serenity as the Sarabhai Crater, an enduring symbol of his impact on space exploration.

In order to forever commemorate his valuable inputs to space exploration, the Vikram Sarabhai Space Centre (VSSC) was founded in Thiruvananthapuram, situated in Kerala. Dedicated to the advancement of solid and liquid propellants for rockets, the VSSC serves as a concrete representation of Sarabhai's unwavering commitment to the progress of science.

Dr. Vikram Sarabhai's indomitable spirit, unwavering vision, and relentless dedication continue to inspire generations of scientists and space enthusiasts in India and beyond. His legacy as a pioneering institution builder and space visionary remains an enduring source of inspiration, propelling India's scientific endeavours to new frontiers of knowledge and exploration. As India continues its journey in space exploration, it does so with the firm foundation laid by the Father of the Indian space program, Dr. Vikram Sarabhai.

INCOSPAR

The Department of Atomic Energy (DAE) was founded in 1954, with Jawaharlal Nehru serving as its first minister and Homi Bhabha as its secretary, with its headquarters in Mumbai, Maharashtra. Under the DAE, the Indian National Committee for Space Research (INCOSPAR) was established in 1962 under the guidance of scientist Vikram Sarabhai, who recognized the importance of space research. It was tasked with formulating India's space program and initially functioned as a part of the Tata Institute of Fundamental Research. Taking over the space science and research responsibilities of the DAE, INCOSPAR was instrumental in setting up the Thumba Equatorial Rocket Launching Station (TERLS) at Thumba in the southern tip of India.

During the 1960s, India made significant strides in the field of space science. The establishment of the Department of Atomic Energy (DAE) in 1950, with Homi Bhabha as its secretary, played a pivotal role in providing funding for space research across the country. This support laid the groundwork for the advancement of space exploration. Throughout the decade, India continued to conduct tests and research on various aspects of meteorology and the Earth's magnetic field, building upon the knowledge accumulated since the inception of the Colaba Observatory in 1823. These studies contributed to a deeper understanding of atmospheric and geomagnetic phenomena. In 1954, the Aryabhatta Research Institute of Observational Sciences (ARIES) was founded in the serene foothills of the Himalayas. ARIES became a significant centre for astronomical research, fostering observational studies and astronomical advancements.

Additionally, the Rangpur Observatory was established in 1957 at Osmania University, Hyderabad. This observatory further expanded India's capabilities in space research, especially in the field of astrophysics and celestial observations. Amidst this growing momentum, the government of India continued to encourage and support space research endeavours. The nation recognized the potential of space exploration as a critical area of scientific and technological development, and investments in this domain were seen as essential for India's progress. Overall, the 1960s marked a crucial phase in India's journey towards becoming a prominent player in space science. The foundation laid during this decade paved the way for future achievements, leading to the establishment of the Indian Space Research Organisation (ISRO) in 1969, which would propel India into the League of spacefaring nations.

Given the limited knowledge of advanced rockets in India, the government and INCOSPAR decided to draw expertise from the Indian Ordnance Factories Service (IOFS) officers, who possessed knowledge of propellants and advanced lightweight materials used in rocket construction. Shri H.G.S. Murthy, an IOFS officer, was appointed the first director of TERLS, marking the commencement of upper atmospheric research in India. The first sounding rocket to launch from Thumba was the American Nike-Apache on November 21, 1963, followed by two-stage rockets from France (Centaure) and Russia (M-100).

A transformative moment took place that would shape the trajectory of India's space exploration journey. B Ramakrishna Rao, Pramod Kale, Prakash Rao, H G S Murthy, D Easwara Das, R Aravamudan and APJ Abdul Kalam; these young and talented scientists and engineers were selected for a momentous opportunity – to undergo specialized training at NASA, the National Aeronautics and Space Administration in the United States. Little did they know that this experience would pave the way for India's ambitious civilian space program, spearheaded by the Indian National Committee for Space Research (INCOSPAR).

The selection of these seven pioneers was a result of the visionary leadership of Dr. Vikram Sarabhai, widely regarded as the Father of the Indian Space Program. Driven by the belief that space technology held the key to addressing the developmental needs of the country, he identified these bright minds to be the future torchbearers of India's space endeavours. The training at NASA exposed young scientists and engineers to cutting-edge technology, scientific methodologies, and advanced research practices. It enriched their knowledge and honed their skills, laying a solid foundation for them to embark on the momentous task of establishing a space research program back home. Upon their return to India, these seven visionaries brought back a wealth of knowledge and expertise, which they used to establish the Indian National Committee for Space Research (INCOSPAR) in 1962.

The pioneers of India's space programme – (From left) R Aravamudan, APJ Abdul Kalam, HGS Murthy, Ramakrishna Rao and D Easwara Das.

India soon ventured into the development of indigenous sounding rockets known as Rohini, which began launching from 1967 onwards. Waman Dattatreya Patwardhan, another IOFS officer, played a pivotal role in developing the propellant for these rockets. Notably, the renowned A. P. J. Abdul Kalam was among the initial team of rocket engineers under INCOSPAR.

The First Rocket

Let us take you back to a fascinating story about India's first rocket launch involving a bicycle and a church. It was the 1960s, a time when most countries, including the USA and the USSR, were gearing up for the space race. During this period, physicist and astronomer Dr. Vikram Sarabhai discovered a small fishing village named Thumba in Thiruvananthapuram. He believed it to be the perfect location for India's inaugural rocket launch. However, there was a slight hitch - the presence of Mary Magdalene Church.

Dr. Sarabhai was of the opinion that Saint Mary Magdalene Church was situated on the Earth's magnetic equator, an imaginary line around the Earth near the equator where the magnetic needle moves from north to south. To proceed with the launch, he and his colleagues approached the then Bishop of Thiruvananthapuram, Reverend Dr. Peter Bernard Pereira, hoping to acquire the church. In an unexpected twist during their discussion, Reverend Pereira asked them to attend Sunday mass. Remarkably, during his sermon at mass, Reverend Pereira addressed the entire issue and the famous scientist's presence, seeking the church and its grounds for space science and research. He emphasized that science seeks truth to enrich human life and questioned if they could give God's abode for a scientific mission. The congregation responded with a heartfelt "Amen," resonating throughout the church and town.

Supported by the whole community, papers were signed, and villagers moved to neighboring areas, even forming a new church to continue their worship. Consequently, right outside Saint Mary Magdalene's Church, India's first rocket launcher, the Thumba Equatorial Rocket Launching Station, was constructed. The church, initially a tiny thatched-roof building built by Saint Francis Xavier, was transformed into a station. Dr. Kalam, one of the scientists along with Sarabhai, recalled the transformations that took place within the church. The prayer room became the first laboratory, the bishop's room turned into a drawing office, and the cattle shed was converted into a lab for the scientists. The main church building was preserved and later transformed into a space museum.

The parts required for the NASA-made rocket, Nike Apache, were so small that they were carried on a bicycle. Finally, on November 21, 1963, India's first rocket was successfully launched into space from the Thumba Equatorial Rocket Launching Station, which would later become the Vikram Sarabhai Space Center, a testament to the collaboration between science and faith. This inspiring tale showcases how determination, unity, and a shared vision can lead to remarkable achievements in the pursuit of scientific progress and space exploration. Below are some of the pictures clicked during the event.

Aravamudan (L) and APJ Abdul Kalam at the Thumba Station in Thiruvananthapuramm, Kerala

Rocket being carried to launch site on Bicycle

Setting up the launch tower and successful launch

Space Science and Technology Centre (SSTC)

The year 1965 marks a significant milestone with the establishment of the Space Science and Technology Centre (SSTC) in a small fishing village named Thumba on 1st January 1965. This pivotal event laid the foundation for the Indian Space Research Organisation (ISRO). Located in the coastal town of Thumba, Kerala, the SSTC was the brainchild of the visionary Dr. Vikram Sarabhai, the founding father of India's space program. Dr. Sarabhai recognized the immense potential of space research and its applications for the country's development. With this vision in mind, the SSTC was established under the umbrella of the Indian National Committee for Space Research (INCOSPAR), the precursor to ISRO.

One of the key factors in choosing Thumba as the location for SSTC was its proximity to the magnetic equator, a unique geographical feature that offered favourable conditions for atmospheric and ionospheric research. This strategic choice allowed scientists to conduct experiments and launch sounding rockets to study the Earth's upper atmosphere easily.

The primary objective of SSTC was to conduct scientific research in space and related fields, laying the groundwork for India's space journey. It served as a centre of excellence for studying cosmic rays, atmospheric science, ionosphere, and other vital aspects of space science. The establishment of SSTC also brought together a team of brilliant minds, including scientists, researchers, and engineers, who passionately worked towards advancing India's space capabilities. Their relentless efforts and dedication would later shape ISRO into the prestigious space agency it is today. Through SSTC, India made its first strides into space research and technology, marking the beginning of a remarkable journey that continues to this day. The centre's pioneering work laid the groundwork for future satellite launches, interplanetary missions, and space explorations that have garnered global acclaim for ISRO.

Satellite Telecommunication Earth Station

Since its inception, the Indian Space Research Organisation (ISRO) has been a well-orchestrated endeavour, comprising three distinct elements that have propelled India's space program to great heights. These elements include satellites for communication and remote sensing, a robust space transportation system, and various application programs. One of the noteworthy milestones in ISRO's journey was the operationalization of the first 'Experimental Satellite Communication Earth Station (ESCES)' on January 01, 1967, in Ahmedabad, Gujarat. This pioneering facility was established in Ahmedabad and served as a trailblazing step in the realm of satellite communication. The ESCES played a dual role, serving as both a vital satellite communication station and a dedicated training centre for Indian and international scientists and engineers. Its significance as a training hub further exemplified India's commitment to nurturing global collaboration and knowledge exchange in space technology.

Birth of ISRO

The significant strides made by INCOSPAR laid the groundwork for the establishment of the Indian Space Research Organisation (ISRO) in 1969. On June 01, 1972, the Space Commission and the Department of Space were established, marking a new era in India's space exploration journey. This significant event streamlined and centralized the nation's space activities, paving the way for ISRO's remarkable achievements in the years to come. The Department of Space took the lead in planning, executing, and overseeing all space-related initiatives and programs in the country, ensuring a well-coordinated approach towards advancements in space science and technology. ISRO was brought under the umbrella of the Department of Space, institutionalizing space research in India and solidifying the Indian space program in its current form. This move was crucial in shaping ISRO's journey, propelling India to become a major player in the global space arena.

As the successor to INCOSPAR, ISRO emerged as an autonomous space agency, reflecting India's dedicated commitment to its space program and the pursuit of cutting-edge space technologies. With the inception of ISRO, India embarked on a transformative journey marked by remarkable achievements in space research, satellite launches, and interplanetary missions. Over the years, ISRO has earned a global reputation, synonymous with India's space excellence, owing to its cost-effective space missions and groundbreaking technologies.

The Satellite Launch Vehicle (SLV)

The SLV was an important project initiated by the Indian Space Research Organisation (ISRO) in the early 1970s to develop satellite launch technology. The primary goal was to enable India to launch satellites into space. Small rockets, such as the SSLV (Small Satellite Launch Vehicle), were also introduced to cater specifically to nano and micro-satellites weighing less than 10 kg and 100 kg, respectively. These small rockets provided on-demand launch services, eliminating the need for clients to wait for larger rockets as co-passengers.

The SLV was designed as a four-stage rocket, employing solid-propellant motors for all stages. Its mission was to reach an altitude of 400 kilometres and carry a payload of 40 kg. The first experimental flight of the SLV occurred in August 1979 but ended in failure. However, ISRO persevered, and on 18th July 1980, the first successful launch of the SLV took place.

All four SLV launches were conducted at the Sriharikota High Altitude Range from the SLV Launch Pad. The first two launches were experimental in nature, while the subsequent two were categorized as developmental launches. The SLV project was significant for India as it marked the country's early foray into space launch capabilities and paved the way for further advancements.

It took approximately seven years from the project's inception to realize the SLV. The first and second-stage solid motor cases were constructed from 15 CDV6 steel sheets, while the third and fourth stages were made from fiber-reinforced plastic.

Despite the challenges and initial setback, the successful development and launches of the SLV showcased India's growing capabilities in space technology. The SLV project played a pivotal role in laying the foundation for future space missions and solidified ISRO's position as a formidable player in the global space community.

Difficult Beginning

In the year 1979, Dr Abdul Kalam found himself at the centre of a momentous mission as the project director for ISRO's SLV-3 mission. The mission was to place a satellite into orbit, and for nearly a decade, thousands of dedicated individuals had worked tirelessly towards this goal. The launch day had arrived, and Dr. Kalam was at Sriharikota, watching as the countdown progressed.

Tension filled the air as the final minutes approached - T minus 4 minutes, T minus 3 minutes, T minus 2 minutes, T minus 1 minute, and finally, T minus 40 seconds. Then, the unexpected happened. The computer detected a problem and put the launch on hold. Dr. Kalam, as the mission director, faced a critical decision. A team of experts analyzed the situation and identified a control issue. The first stage of the launch went smoothly, but the second stage encountered a problem, causing the rocket to spin out of control. Instead of reaching orbit, the satellite found itself in the Bay of Bengal.

It was the first time in his career that Dr. Kalam faced failure, and he questioned how to manage it. Despite the setback, he made the courageous decision to overrule the computer warning and launch the system manually. The mission, however, did not succeed as hoped. In the aftermath, the then ISRO chief, Satish Dhawan, shouldered the blame, taking full responsibility for the failure. He stood before the world and said, "Dear friends, we have failed today. I want to support my technologists, my scientists, my staff so that next year they succeed." Despite criticism, his unwavering support for the team left a lasting impact on Dr. Kalam.

The following year, on July 18, 1980, the Rohini RS-1 satellite was successfully launched into orbit. Satish Dhawan asked Dr. Kalam to hold a press conference to share the success. This experience taught Dr. Kalam an essential lesson in leadership and management - owning failure and attributing success to the team. It was a lesson that came from experience, shaping his approach to leadership throughout his life.

Dr. Abdul Kalam's story is one of resilience, determination, and the power of strong leadership. It serves as an inspiration to generations to come, reminding us that in both success and failure, great leaders stand with their teams, guiding them towards new heights in the pursuit of exploration and discovery.

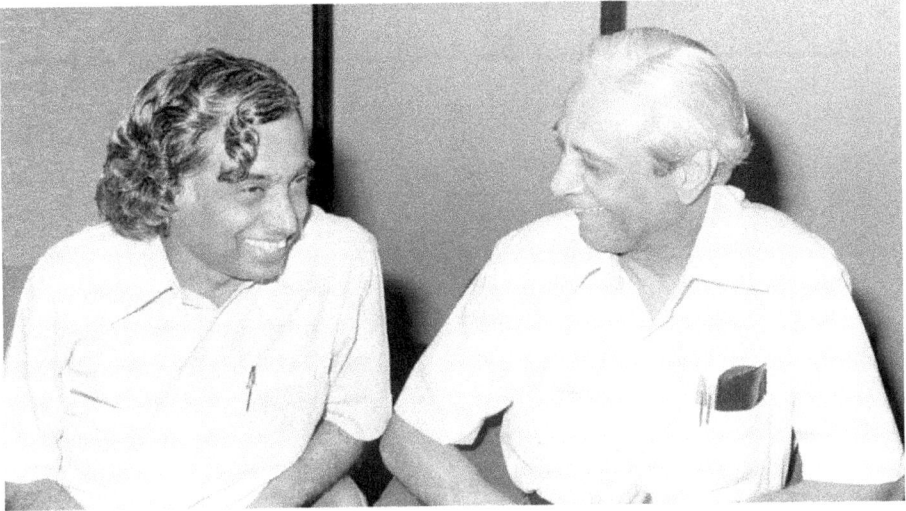

APJ Abdul Kalam (L) with rocket scientist Prof Satish Dhawan

First Indian Satellite:

Aryabhata

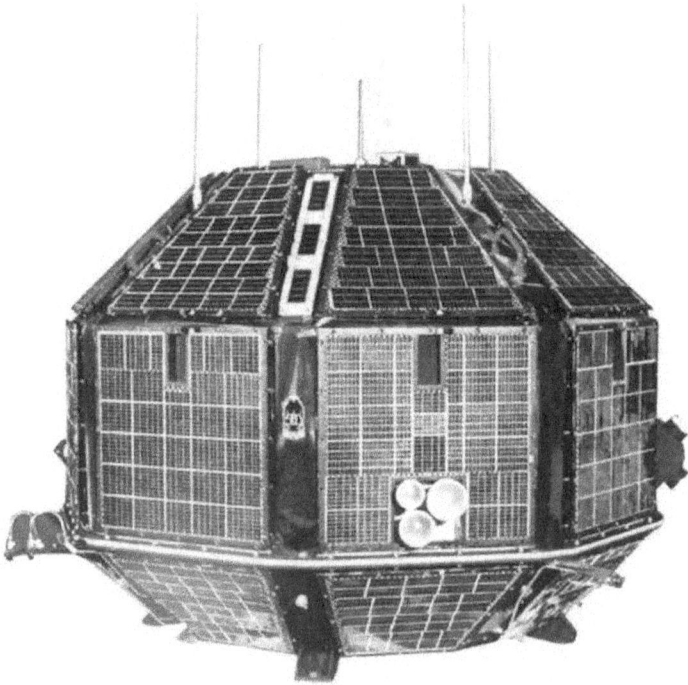

India's first satellite, Aryabhata, derived its name from the renowned mathematician-astronomer Aryabhata, who flourished during the classical age of Indian mathematics and astronomy in the 5th century. His notable works, the Āryabhaṭīya and the Arya-siddhanta, contributed significantly to ancient Indian knowledge in these fields. In 1972, Dr. Vikram Sarabhai entrusted one of his brilliant scientists, U R Rao, with the ambitious task of creating an indigenous Indian satellite that would orbit the Earth.

At the time, U R Rao was the only Indian who had participated in two NASA satellite projects in the United States. He assembled a team of young talents from the Indian Institute of Science in Bangalore to work on the project. Within a remarkable span of just 30 months and a budget of 3.5 crore rupees, they completed the spacecraft, which took the shape of a 26-sided polyhedron with a diameter of 1.4 meters. The surfaces, except for the top and bottom, were covered with solar cells, and the satellite carried three payloads. When Prime Minister Indira Gandhi visited the project for inspection, U R Rao presented her with three choices for the satellite's name: Maitri, Jawahar, and Aryabhata. Ultimately, they settled on Aryabhata, as Rao explained to Indira Gandhi the significant contributions made by the ancient Indian mathematician, including his explanation of solar and lunar eclipses and his calculation of the value of pi up to four decimal places.

Aryabhata was assembled near Bangalore and launched by India on 19 April 1975 from Kapustin Yar, a Russian rocket launch and development site in Astrakhan Oblast, using a Kosmos-3M rocket. The satellite's primary mission was to explore the conditions in the Earth's ionosphere, measure neutrons and gamma rays from the sun, and conduct investigations in X-ray astronomy. The historic event of its launch was commemorated on two rupee notes and postal stamps by both India and the Soviet Union. Regrettably, a power failure led to Aryabhata operating for only five days in space before all communication was lost. Nevertheless, despite this setback, the mission served as a stepping stone for ISRO to evolve into one of the world's most cost-effective and successful space organizations. Aryabhata's journey paved the way for India's remarkable progress in space research, satellite launches, and interplanetary missions, making it a key player in the global space exploration.

Satellite Instructional
Television Experiment (SITE)

In 1975-76, India embarked on a groundbreaking endeavour known as the Satellite Instructional Television Experiment (SITE), often regarded as 'the largest sociological experiment in the world.' This innovative project aimed to extend developmental programs to rural communities using the American Technology Satellite (ATS-6). Over 200,000 people across 2400 villages in six states benefited from this experiment, which played a pivotal role in educating both educators and learners.

SITE marked India's initial step towards integrating technology into education. Seeking to bridge the communication gap in rural areas, the experiment utilized American technology through collaboration between NASA and the Indian Space Research Organization (ISRO). It was an ambitious venture, emphasizing the potential of satellite broadcasting in bringing education to underserved communities. The project's objectives included offering educational content to economically disadvantaged and illiterate segments of society while also nurturing India's proficiency in satellite communications.

Running from August 1, 1975, to July 31, 1976, SITE covered more than 2400 villages spanning 20 districts across six Indian states. All India Radio produced educational television programs, which were beamed to these areas via NASA's ATS-6 satellite. The project garnered international support from organizations like UNDP, UNESCO, UNICEF, and ITU, underlining its global significance. The success of SITE extended beyond its educational impact. It played a pivotal role in developing India's satellite program, INSAT, showcasing the country's technological prowess. This pioneering experiment demonstrated India's ability to address socioeconomic needs through cutting-edge technology.

In essence, the Satellite Instructional Television Experiment brought educational content to rural India and laid the foundation for India's advancement in space technology and its commitment to empowering all segments of society.

Second Indian Satellite:

Bhaskara 1 & 2

Bhaskara-I and Bhaskara-II were two significant satellites developed by the Indian Space Research Organisation (ISRO), serving as India's first low-Earth orbit Earth observation satellites. Both satellites were named after ancient Indian mathematicians, Bhāskara I and Bhāskara II, paying tribute to their contributions to the field of mathematics.

Bhaskara-I, weighing 444 kg at launch, was successfully launched from Kapustin Yar on June 7, 1979, using the C-1 Intercosmos Launch Vehicle. It was placed in an orbital path with a perigee and apogee of 394 km and 399 km, respectively, at an inclination of 50.7°. The satellite was equipped with advanced instruments, including two television cameras operating in the visible and near-infrared spectrum, enabling data collection related to hydrology, forestry, and geology. Additionally, it featured a Satellite Microwave Radiometer (SAMIR) operating at 19 and 22 GHz, which facilitated the study of ocean-state, water vapour, and liquid water content in the atmosphere. The satellite also boasted an X-ray sky monitor operating in the 2-10 keV energy range, making it possible to detect transient X-ray sources and monitor long-term spectral and intensity changes in the X-ray sources.

Bhaskara-II, the second satellite in the series, was launched on November 20, 1981, from the Volgograd Launch Station using the C-1 Intercosmos Launch Vehicle. This satellite was crucial in providing valuable data on ocean and land surface conditions. Despite experiencing a malfunction in one of its two onboard cameras, Bhaskara-II managed to transmit more than two thousand images before its re-entry in 1991. It orbited at an altitude of 541 × 557 km with an inclination of 50.7°, contributing significantly to India's growing capabilities in Earth observation.

The successful missions of Bhaskara-I and Bhaskara-II showcased ISRO's expertise in satellite technology and data collection, cementing India's position as a formidable player in the realm of space-based Earth observation. These satellites continue to hold historical importance as they laid the groundwork for future advancements in the field of remote sensing and scientific research.

India's first Experimental Satellite Vehicle (SLV-3)

In India's space exploration journey, a significant milestone was achieved with the successful launch of the Satellite Launch Vehicle-3 (SLV-3) on July 18, 1980, from the Sriharikota Range (SHAR). SLV-3 marked India's first experimental satellite launch vehicle, making the country the sixth nation in the world to possess space-faring capabilities.

SLV-3, an all-solid, four-stage vehicle weighing 17 tonnes and standing at a height of 22 meters, was designed to place 40 kg class payloads in Low Earth Orbit (LEO). The vehicle's inaugural launch in August 1979 carried a Rohini technology payload but was only partially successful as it failed to inject the satellite into its desired orbit. However, determination and continued efforts led to the successful launch in July 1980, where the Rohini Series-I satellite, RS-1, was placed in orbit, marking India's entry into an exclusive group of space-faring nations.

The SLV-3 launch vehicle employed an open-loop guidance system with a stored pitch program to steer the vehicle along a pre-determined trajectory during its flight. This proved to be a critical stepping stone for India's space program, providing valuable insights and learnings for more advanced launch vehicle projects that followed, such as the Augmented Satellite Launch Vehicle (ASLV), the Polar Satellite Launch Vehicle (PSLV), and the Geosynchronous Satellite Launch Vehicle (GSLV).

In addition to the successful July 1980 launch, SLV-3 launched two more in May 1981 and April 1983, both orbiting Rohini satellites equipped with remote sensing sensors. These missions further strengthened ISRO's capabilities and laid the foundation for future advancements in space technology.

The successful culmination of the SLV-3 project marked the beginning of a new era for India's space program, propelling the nation towards achieving greater heights in satellite technology and space exploration. It showcased India's prowess in space research and set the stage for more ambitious and successful space missions in the years to come.

Vikram Sarabhai (C), APJ Abdul Kalam (to Sarabhai's left), G Madhavan Nair (former Isro chief), C R Satya and HGS Murthy inspecting the manufacturing process.

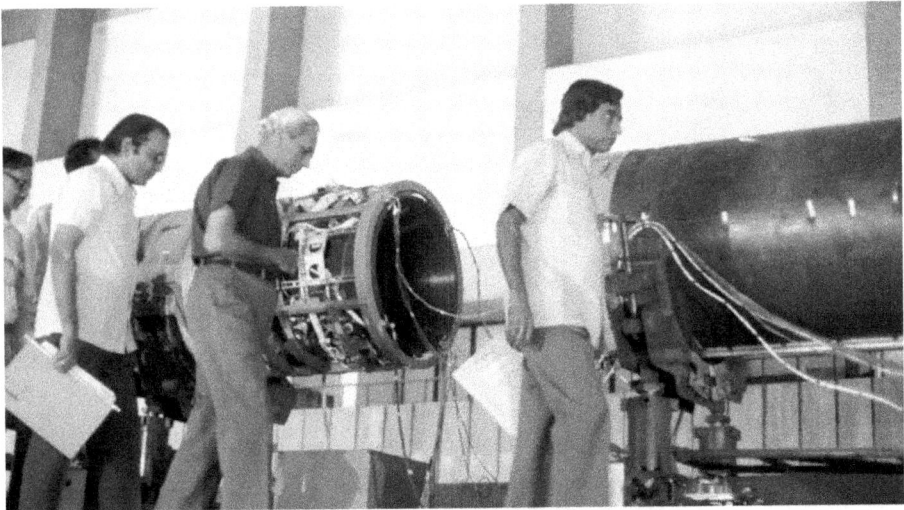

APJ Abdul Kalam and Prof Satish Dhawan and others inspecting the SLV-III integration facility.

In July 1980, Dr. APJ Abdul Kalam addressed the audience, including Prof. Satish Dhawan and other dignitaries on the dais, to commemorate the successful launch of SLV-III, India's inaugural experimental satellite launch vehicle.

Dr. APJ Abdul Kalam

There are some who question the relevance of space activities in a developing nation. To us, there is no ambiguity of purpose. We do not have the fantasy of competing with economically advanced nations in the exploration of the Moon or the planets or manned space-flights. But we are convinced that if we are to play a meaningful role nationally, and in the community of nations, we must be second to none in the application of advanced technologies to address the real problems of man and society.

These are the words of India's beloved Dr. Abdul J Kalam, a pivotal figure in the Indian Space program. He was born on October 15, 1931, in Rameswaram, Tamil Nadu. His legacy resonates not only in the realm of space exploration but also as an inspiration, with his birthday observed as "World Students Day."

After graduating from the Madras Institute of Technology in 1960, Kalam began his journey by joining the Aeronautical Development Establishment of the Defence Research and Development Organisation (DRDO). Although he initially focused on designing a small hovercraft, he remained uncertain about his role at DRDO. Under the mentorship of Vikram Sarabhai, the revered space scientist, Kalam became part of the INCOSPAR committee. This affiliation paved the way for his transition to the Indian Space Research Organisation (ISRO) in 1969.

Within ISRO, Kalam's influence burgeoned as he assumed the role of project director for India's maiden Satellite Launch Vehicle (SLV-III) initiative. In a remarkable achievement in July 1980, the SLV-III successfully deployed the Rohini satellite into near-earth orbit. This accomplishment culminated Kalam's earlier independent work on an expandable rocket project during his tenure at DRDO. His dedication led to the expansion of this program, marked by government approval and the inclusion of additional engineers.

During his career, Kalam's engagement extended beyond national boundaries. Notably, he visited esteemed institutions such as NASA's Langley Research Center, Goddard Space Flight Center, and Wallops Flight Facility, contributing to his scientific enrichment.

Kalam's association with Vikram Sarabhai, known as the Father of the Indian Space Program, deeply influenced his trajectory. He was instrumental in launching India's first sounding rocket on November 21, 1963. Kalam's transformational period occurred while serving at ISRO, where he harnessed the momentum to lead groundbreaking projects, particularly the development of the Polar Satellite Launch Vehicle (PSLV) and SLV-III.

Interestingly, while Kalam'sreputation as "India's Missile Man" remains prominent due to hiscontributions to the defense sector, his achievements in the field of spaceexploration, encapsulated by the moniker "India's Rocket Man," areequally deserving of recognition. His leadership extended beyond ISRO, notablysteering the government's Integrated Guided Missile Development Programme.Kalam's diverse accomplishmentsculminated in his election as the 11th President of India in 2002. His journeyreflects not only scientific excellence but also a steadfast commitment toadvancing India's capabilities in space technology and beyond. The impact ofhis work resonates not only within the scientific community but also in thebroader social and cultural sphere, leaving an indelible mark on the nation'spursuit of excellence in space exploration.

APPLE Satellite

The Ariane Passenger Payload Experiment (APPLE) was a groundbreaking achievement for the Indian Space Research Organisation (ISRO) and marked India's foray into experimental communication satellites. Launched on June 19, 1981, aboard the European Space Agency's (ESA) Ariane launch vehicle from Centre Spatial Guyanais in French Guiana, APPLE became ISRO's first indigenous communication satellite.

APPLE was designed and built in just two years, showcasing ISRO's ingenuity and capability to achieve remarkable milestones with limited infrastructure in industrial sheds. It was launched into Geosynchronous Transfer Orbit (GTO) by ESA's Ariane vehicle, and subsequently, ISRO's apogee motor derived from the fourth stage of SLV-3 boosted it into Geosynchronous Orbit (GEO).

Serving as a testbed for India's telecommunications space relay infrastructure, APPLE provided valuable hands-on experience in designing and developing three-axis stabilized geostationary communication satellites. It offered insights into crucial aspects of space missions, including in-orbit deployment of appendages, station keeping, and orbit-raising manoeuvres.

Throughout its mission, APPLE was extensively utilized to conduct experiments on various communication systems, such as time, frequency, and code division multiple access systems. It also played a crucial role in radio networking, computer interconnect, and pocket switching experiments. Despite facing challenges like the failure of one solar panel deployment, APPLE successfully relayed TV programs. It facilitated radio networking, demonstrating its effectiveness as India's first three-axis stabilized experimental Geostationary communication satellite.

APPLE's mission contributed to developing India's communication infrastructure and served as a platform to introduce state-of-the-art technologies. It incorporated momentum-biased three-axis stabilization techniques, motor-driven deployed solar arrays, earth sensing for attitude control, and C-band transponder design, among other advancements. Overall, APPLE's success showcased ISRO's commitment to technological excellence and laid the groundwork for future advancements in the field of space communication and satellite technology.

Satellite on Bullock Cart

In 1981, India's space scientists faced an extraordinary challenge as they prepared to launch their first communication satellite, APPLE, from the Guiana Space Centre in France. The momentous mission was a pioneering step in India's space program but encountered setbacks on multiple fronts.

During those times, communication options were limited, with only telephones and telex networks available for domestic and international communication. The Indian Space Research Organisation (ISRO) did not possess the advanced technology and infrastructure available today. Setting up the APPLE Mission Control Centre at Sriharikota presented significant challenges.

One crucial aspect that needed attention was testing the satellite's antenna and ensuring proper Telemetry, tracking, and control (TT&C) links to maintain communication with the satellite in space. However, conducting this test required a specialized facility with the satellite structure placed under a thermal blanket, something ISRO needed to gain four decades ago.

Faced with this dilemma, the inventive minds at ISRO devised an unexpected solution - they loaded the satellite onto a bullock cart! This seemingly unconventional approach served a crucial purpose. The bullock cart provided a non-magnetic environment, allowing the engineers to conduct the antenna test in an open field to address the impedance matching problem affecting the TT&C link.

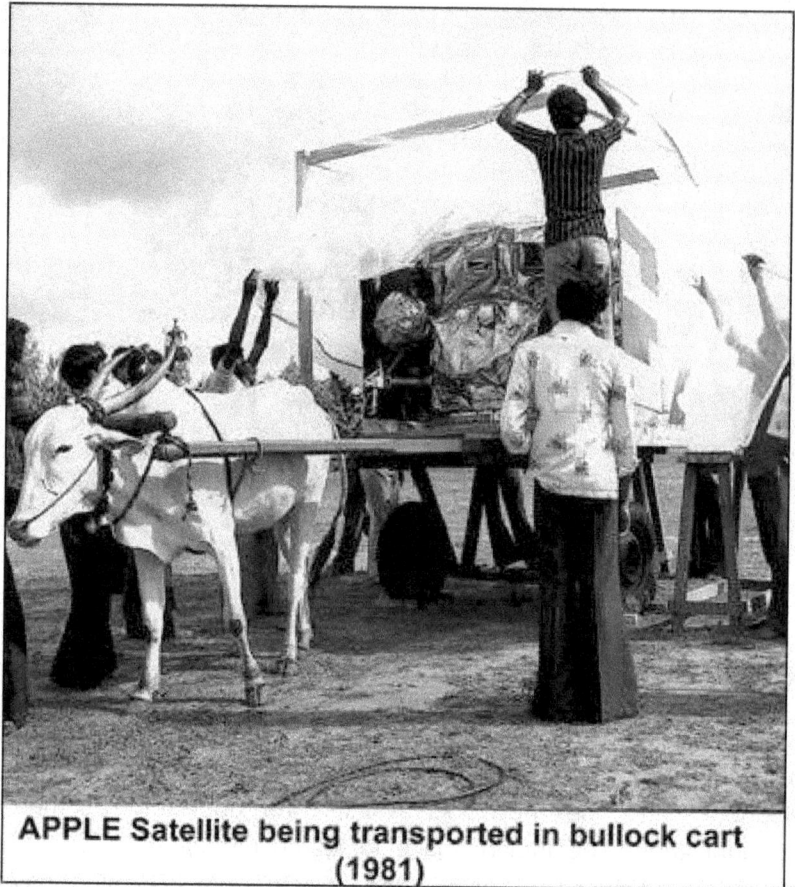

APPLE Satellite being transported in bullock cart (1981)

The decision to utilize a humble bullock cart for such a critical scientific mission highlights the determination and resourcefulness of ISRO's early scientists. They overcame hurdles with limited resources, ensuring that APPLE's mission continued on track without needing expensive and time-consuming alternatives.

And so, the rocket science behind the bullock cart became an integral part of India's space story, symbolizing the ingenuity and spirit of innovation that defined ISRO's journey into space exploration. This remarkable endeavour laid the foundation for indigenous satellite development, ultimately shaping India's space program into today's extraordinary force.

Indian National Satellite System (INSAT)

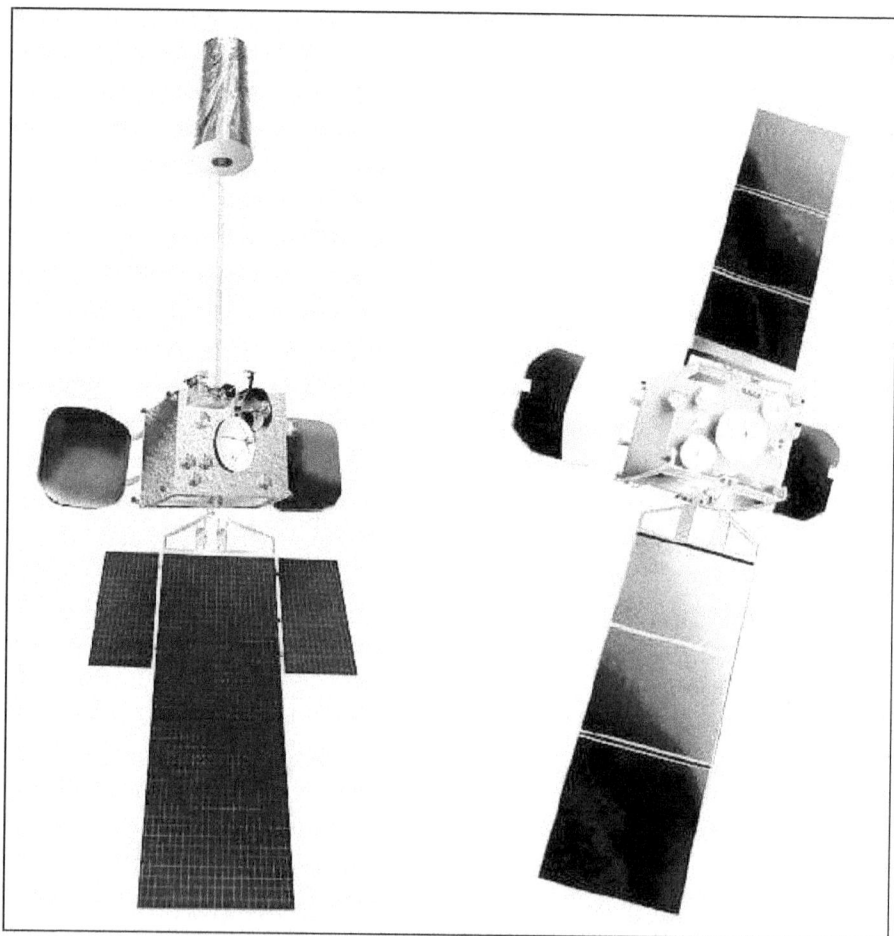

The Indian National Satellite System (INSAT) represents a vital series of geostationary satellites launched by the Indian Space Research Organisation (ISRO), serving multiple purposes like telecommunications, broadcasting, meteorology, and search and rescue operations. Established in 1983, INSAT stands as the most extensive domestic communication system in the Indo-Pacific Region. The system is a joint venture involving prominent Indian government agencies such as the Department of Space, Department of Telecommunications, India Meteorological Department, All India Radio, and Doordarshan. The overall coordination and management of the INSAT system rests with the Secretary-level INSAT Coordination Committee.

The primary objective of INSAT satellites is to provide transponders in various bands to cater to India's television and communication needs. Some satellites in the INSAT series are equipped with Very High Resolution Radiometer (VHRR) and Charge Coupled Device (CCD) cameras, enabling meteorological imaging. Moreover, these satellites also incorporate transponders for receiving distress alert signals, facilitating search and rescue missions in the South Asian and Indian Ocean Region, as ISRO participates in the Cospas-Sarsat program.

The INSAT program commenced with the launch of INSAT-1B in August 1983. The first satellite, INSAT-1A, launched in April 1982, couldn't fulfil its mission objectives. INSAT-1B ushered in a revolution in India's television and radio broadcasting, telecommunications, and meteorological sectors. It facilitated the rapid expansion of TV and modern telecommunication facilities even to remote areas and offshore islands. The INSAT system provides transponders in C, Extended C, and Ku bands, catering to diverse communication services. Certain INSAT satellites also carry instruments for meteorological observation and data relay to provide essential meteorological services. Notably, KALPANA-1 serves as an exclusive meteorological satellite. The operations of these satellites are monitored and controlled by Master Control Facilities located in Hassan and Bhopal.

Out of the 24 satellites launched in the INSAT program, 11 satellites are still operational. One is INSAT-2E, the last of the six satellites in the INSAT-2 series. INSAT-2E carries seventeen C-band and lower extended C-band transponders, providing zonal and global coverage with an Effective Isotropic Radiated Power (EIRP) of 36 dBW. Additionally, it is equipped with VHRR, offering imaging capabilities in the visible, thermal infrared, and water vapour channels, with ground resolutions of 2x2 km and 8x8 km.

Another noteworthy satellite is INSAT-3A, a multipurpose satellite launched by Ariane in April 2003, located at 93.5 degrees East longitude. It carries 12

Normal C-band transponders and 6 Ku-band transponders, providing communication coverage to regions from the Middle East to South East Asia. INSAT-3A also includes VHRR and CCD camera payloads for imaging and data relay transponders for global receive coverage.

INSAT-3C, launched in January 2002, is positioned at 74 degrees East longitude and provides a combination of 24 Normal C-band transponders and six Extended C-band transponders, offering extensive coverage over India. Similarly, INSAT-3D, launched in July 2013, operates at 82 degrees East longitude and covers a large part of the Indian Ocean region, rendering distress alert services to countries like India, Bangladesh, Bhutan, Maldives, Nepal, Seychelles, Sri Lanka, and Tanzania.

INSAT-3DR, launched on 9 September 2016 by the GSLV Mk II F05, is a dedicated weather satellite with a 6-channel imager and a 19-channel sounder, serving India's meteorological requirements.

On the other hand, INSAT-3E, launched in September 2003, was positioned at 55 degrees East longitude and provided Normal C-band and Extended C-band transponders for coverage over India. However, the satellite has been decommissioned and out of service since April 2014, with GSAT-16 set to replace it.

The INSAT-4 series includes INSAT-4A, launched in December 2005, with both Ku-band and C-band transponders providing extensive coverage over India. INSAT-4B, launched in March 2007, carries similar payloads as INSAT-4A, augmenting high power transponder capacity over India in the Ku and C bands.

INSAT-4CR, launched on 2 September 2007, serves as a replacement for the lost INSAT-4C satellite. It offers 12 Ku-band transponders for communication purposes and includes a Ku-band Beacon for satellite tracking. The satellite is designed to operate for ten years.

The INSAT system stands as a testament to ISRO's prowess in space technology and its dedication to advancing applications for the benefit of India and the Indo-Pacific Region. These satellites have been instrumental in enhancing communication, broadcasting, and meteorological services while playing a crucial role in search and rescue missions. The success and continued operation of these satellites.

INSAT-1A Satellite:

On April 10, 1982, India's ambitious INSAT-1A satellite, a vital part of the Indian National Satellite System, was launched into orbit with the assistance of Ford Aerospace and NASA. This cutting-edge satellite was equipped with 12 C and 3 S-band transponders and had a mission duration of seven years, aimed at revolutionising communications and weather observation in the country.

However, shortly after its successful launch, INSAT-1A faced several challenges. The deployment of its antennas, solar array, and stabilisation boom proved problematic. The C-band antenna remained undeployed for almost two weeks, and the solar array couldn't extend fully, impacting weather observation capabilities. Additionally, the S-band transponders overheated and ceased functioning.

The team confronted further issues when they detected a fault in the attitude control system, leading to a faster propellant depletion. Despite their efforts, the stabilisation boom failed to deploy, leaving INSAT-1A in a precarious state.

On September 4, 1982, an attempt was made to temporarily deactivate the primary Earth-tracking sensor to mitigate the Sun's interference. Unfortunately, due to the boom malfunction, the backup sensor was oriented towards the Moon and eventually shut down under the strain. Efforts to regain Earth-lock depleted the satellite's propellant reserves, leaving it stranded in space. Consequently, INSAT-1A's mission concluded in September 1982, less than six months into its intended seven-year venture.

Stretched Rohini Satellite Series

The Stretched Rohini Satellite Series (SROSS) marked a significant step in India's space exploration efforts, developed by the Indian Space Research Organisation (ISRO) as a continuation of the Rohini Satellites. These satellites served various purposes, from astrophysics and Earth remote sensing to upper atmospheric monitoring and innovative application-oriented missions. The SROSS satellites were carried into space aboard the developmental flights of the Augmented Satellite Launch Vehicle.

The first satellite in the series, SROSS A, was launched on 24 March 1987. However, the mission faced a setback as the launch vehicle encountered a failure, preventing SROSS A from reaching its intended orbit. Despite this, the satellite was equipped with two retro-reflectors for laser tracking, showcasing ISRO's dedication to advancing space technology.

In a subsequent attempt, SROSS B was launched on 13 July 1988. Unfortunately, fate repeated itself, and the launch vehicle experienced another failure, resulting in SROSS B failing to reach its designated orbit. SROSS B carried two essential instruments this time: a Monocular Electro-

Optical Stereo Scanner (MEOSS) from West Germany and ISRO's 20-3000keV Gamma-ray Burst Experiment (GRB).

Though both missions did not achieve their planned orbits, the SROSS series represented ISRO's pioneering spirit and determination to explore new frontiers in space. Despite the challenges, these missions provided valuable insights and learnings, which contributed to the growth and development of India's space program. The SROSS project paved the way for future successes and demonstrated ISRO's commitment to scientific research and advancements in space technology.

After the failure of the first two Satellites in the Rohini Series, on 20 May 1992, SROSS 3 achieved a lower-than-planned orbit. It carried out the Gamma-Ray Burst (GRB) experiment, monitoring celestial gamma-ray bursts in the 20–3000 keV energy range. The GRB experiment operated from 25 May 1992 until its re-entry on 14 July 1992, utilizing a primary and redundant CsI(Na) scintillator to record data. A 'burst mode' was triggered by specific count rate thresholds, capturing temporal and spectral data before and after the trigger.

Additionally, SROSS 3 carried the Retarded Potential Analyzer (RPA) experiment to study temperature, density, and electron characteristics in Earth's ionosphere. The onboard computer system employed the RCA CDP1802 microprocessor.

On 4 May 1994, SROSS C2 was launched, featuring improved GRB experiments over its predecessor, SROSS 3. Enhancements to the onboard memory and background spectra measurement after a burst event led to the discovery of twelve candidate events up to 15 February 1995, out of 993 triggers. SROSS 3 and SROSS C2 contributed valuable data to India's space research efforts.

Rohini Satellite RS-D2

Despite the setback with the INSAT-1A satellite, ISRO remained undeterred. Within a year, they achieved a significant milestone with the launch of the

Rohini Satellite RS-D2t. This experimental spin-stabilized satellite weighed 41.5 kg and boasted a power handling capacity of 16W. On April 17, 1983, it was successfully launched onboard the SLV-3 from the SHAR Centre.

The primary payload onboard RS-D2t was the Smart Sensor Camera, which played a crucial role in the mission. This advanced camera captured and transmitted over 5,000 picture frames in visible and infrared bands, allowing for detailed identification of various features. Moreover, it demonstrated the innovative technique of determining attitude and orbit using images.

The camera was equipped with onboard processing capability to classify ground features such as water bodies, vegetation, bare land, clouds, and snow. This capability provided valuable insights for scientific research and further advancements in remote sensing.

The successful launch and operation of RS-D2t marked another significant step forward in India's space exploration journey. The dedication and perseverance of ISRO's scientists and engineers shone through in this mission, reaffirming the organization's commitment to pushing the boundaries of space technology.

First Indian in Space

Rakesh Sharma With Ship Commander Yury Malyshev, right, and Flight Engineer Gennady Strekalov, left.

In 1984, a significant chapter in India's space exploration unfolded when the Indian Space Research Organisation (ISRO) collaborated with the Soviet Union for a manned space mission. The mission, known as "India-Soviet manned space mission," aimed to send an Indian astronaut into space aboard a Soviet spacecraft.

On April 2, 1984, Squadron Leader Rakesh Sharma achieved a historic milestone by becoming the first Indian to journey into space. Flying aboard the Soyuz T11 spacecraft, he spent seven days, 21 hours, and 40 minutes aboard the Salyut 7 orbital station alongside two Soviet astronauts, Yury Malyshev and Gennadi Strekalov. Soyuz T-11 marked the sixth expedition to the Soviet Salyut 7 space station, serving as the platform for this groundbreaking mission that captured the imagination of people worldwide.

The mission showcased India's growing capabilities in space research and solidified its position as a space-faring nation. It also marked a milestone in international collaboration, highlighting the friendly ties between India and the Soviet Union.

While aboard Salyut 7, Rakesh Sharma conducted an Earth observation program focusing on India. He meticulously carried out life sciences and materials processing experiments, including tests on silicon fusing. Remarkably, he also delved into practising yoga to counteract the effects of prolonged orbital spaceflight. After his return to India, Mr. Rakesh Sharma received India's highest peacetime military decoration award, the Ashoka Chakra, on May 7, 1985.

The India-Soviet manned space mission was a significant achievement for ISRO and paved the way for future endeavours in space exploration. Rakesh Sharma's remarkable journey inspires generations of space enthusiasts and remains a proud moment in India's space history.

Augmented Satellite Launch Vehicle (ASLV)

The Augmented Satellite Launch Vehicle (ASLV), also known as the Advanced Satellite Launch Vehicle, was a significant project undertaken by the ISRO during the early 1980s. Its purpose was to develop the necessary technologies for launching payloads into Low Earth orbit (LEO), focusing on enhancing the payload capability for LEO.

The ASLV was designed as a five-stage solid-fuel rocket, building upon the experience gained from the earlier Satellite Launch Vehicle (SLV) missions. The vehicle's primary objective was to demonstrate and validate critical technologies vital for future launch vehicles. These included strap-on technology, inertial navigation, bulbous heat shields, vertical integration, and closed-loop guidance systems.

The ASLV programme witnessed four developmental flights. The first two, carried out on March 24, 1987, and July 13, 1988, were unsuccessful. However, the third developmental flight, ASLV-D3, conducted on May 20, 1992, succeeded by placing the SROSS-C satellite, weighing 106 kg, into an orbit of 255 x 430 km. Subsequently, the fourth and final developmental flight, ASLV-D4, launched on May 4, 1994, also succeeded by placing SROSS-C2, weighing 106 kg, into the intended orbit. SROSS-C2 was equipped with two payloads, the Gamma Ray Burst (GRB) Experiment and the Retarding Potentio Analyser (RPA), and functioned effectively for seven years.

The ASLV project aimed to enhance the payload capability to 150 kg, three times that of SLV3, making it suitable for LEO missions. It was a low-cost intermediate vehicle to bridge the gap between earlier launch vehicles and more advanced ones for future space missions. By conducting the ASLV missions, ISRO sought to pave the way for increased efficiency and reliability in launching satellites into space.

ASLV Rocket Design:

The ASLV rocket was developed using lessons from previous SLV rockets. It was designed as a small-lift launch vehicle featuring a five-stage solid-fuel rocket configuration. Each stage played a crucial role in the vehicle's performance, contributing to the overall thrust and achieving its mission objectives.

The configuration of ASLV consisted of two strap-on boosters acting as the first stage, with the core stage igniting after the booster burnout. The lift-off weight of ASLV was approximately 40 tons, and its height measured 24 meters. The strap-on stage featured two identical solid propellant motors with a 1-meter diameter. This setup enabled the ASLV to have a payload capability of 150 kg class satellites, positioning them into 400 km circular orbits in LEO.

The first stage of the ASLV was equipped with two solid propellants, providing it with a maximum thrust of 502.6 kilo-Newton. The burn duration time for this stage was 49 seconds, during which it provided the initial thrust needed for lift-off.

Following the first stage, the second stage took over the propulsion with a maximum thrust of 702.6 kN. Powered by a single solid propellant, the second stage's burn duration lasted 45 seconds, propelling the rocket further towards its designated trajectory.

Upon completing the second stage's role, the third stage came into action, featuring a maximum thrust of 304 kN. Similar to the previous stage, a single solid-propellant powered it; the third stage's burn duration was 36 seconds, contributing to the ASLV's continuous acceleration.

The fourth stage of the ASLV played a critical part in fine-tuning the rocket's trajectory. With a maximum thrust of 90.7 kN (20,400 lbf) and powered by a single solid propellant, the burn duration for the fourth stage also lasted 45 seconds, further refining the vehicle's flight path.

The final stage, known as the fifth stage, marked the last phase of the launch sequence. Two solid propellants powered it, providing a maximum thrust of 35 kN. It ensured efficient consumption of propellant. The fifth stage's burn duration spanned 33 seconds, culminating in the successful delivery of the payload to its intended orbit.

The ASLV's configuration as a five-stage, all-solid propellant vehicle allowed it to achieve a payload capability of 150 kg class satellites into a 400 km circular orbit. This enhanced payload capability, three times that of its predecessor, the SLV3, made it a significant milestone in India's space exploration journey.

The ASLV project marked India's determination to strengthen its position in the global space community and showcased ISRO's commitment to mastering indigenous rocket technologies. The successful launches of the ASLV-D3 and ASLV-D4 missions, placing SROSS-C and SROSS-C2 satellites into their intended orbits, further demonstrated India's growing capabilities in space technology. While the ASLV project faced initial challenges and budgetary constraints, its achievements laid the foundation for subsequent launch vehicle developments. The success of the ASLV bolstered ISRO's confidence in pursuing more ambitious projects, such as the Polar Satellite Launch Vehicle (PSLV) and the Geosynchronous Satellite Launch Vehicle (GSLV) series.

Despite its achievements, the ASLV programme faced challenges due to budgetary constraints and the simultaneous development of the Polar Satellite Launch Vehicle (PSLV) programme. As a result, the ASLV programme was terminated after the initial developmental flights, with its learnings contributing to subsequent advancements in space technology.

Organization Structure of ISRO

The Department of Space (DoS) is a branch of the Indian government responsible for overseeing the Indian space program. Its main aim is to develop and apply space science and technology to benefit the country's socio-economic growth. Its primary objective is to utilize space science and technology to support India's comprehensive development.

ISRO logo:

In its early years, the (ISRO did not possess an official logo. It was in the year 2002 that they formally adopted a distinctive emblem to represent the organization. The logo design features a prominent orange arrow, symbolizing progress and advancement, shooting upwards towards the future.

This upward-pointing arrow is flanked by two satellite panels coloured in blue, representing the essential aspect of space technology and exploration. One panel is on the arrow's left side, while the other occupies the right side.

The organisation's name, "ISRO," is skillfully integrated into the logo. On the left, the name is presented in the orange-coloured Devanagari script, showcasing the deep-rooted Indian cultural heritage. On the right, the name appears in blue-coloured English letters, precisely rendered in the Prakrta typeface, which adds a touch of modernity and dynamism to the overall design.

The Space Commission is responsible for shaping policies and supervising the execution of India's space program. It aims to foster the growth and application of space science and technology for the nation's socio-economic advancement. The Department of Space (DOS) carries out these programs primarily through entities such as the Indian Space Research Organisation (ISRO), Physical Research Laboratory (PRL), National Atmospheric Research Laboratory (NARL), North Eastern-Space Applications Centre (NE-SAC), and Semi-Conductor Laboratory (SCL). Antrix Corporation, founded in 1992 as a government-owned company, handles space products and services marketing.

National committees oversee the establishment of space systems and their applications. These include the INSAT Coordination Committee (ICC), the Planning Committee on the National Natural Resources Management System (PC-NNRMS), and the Advisory Committee on Space Sciences (ADCOS).

DOS's Secretariat and ISRO's Headquarters are situated at Antariksh Bhavan in Bangalore. ISRO's program offices, located at its headquarters, manage various programs like satellite communication, earth observation, launch vehicles, space science, disaster management support, sponsored research, contracts, international collaboration, safety, reliability, publications, public relations, budget analysis, civil engineering, and human resources development. These entities collectively work towards advancing India's space capabilities and applying space technology for the betterment of the nation.

To achieve these objectives, the Department of Space has formulated various programs:

- The Launch Vehicle program focuses on indigenous capabilities for launching spacecraft.
- The INSAT Program serves telecommunications, broadcasting, meteorology, and educational development.
- The Remote Sensing Program, utilising satellite imagery for multiple developmental purposes.
- Research and Development in Space Sciences and Technology to drive national progress.

DoS oversees various agencies and institutes related to space exploration and technology. These include significant satellite systems like INSAT for communication, broadcasting, and meteorology and the Indian Remote Sensing Satellites (IRS) system for resource monitoring.

Organisation Structure

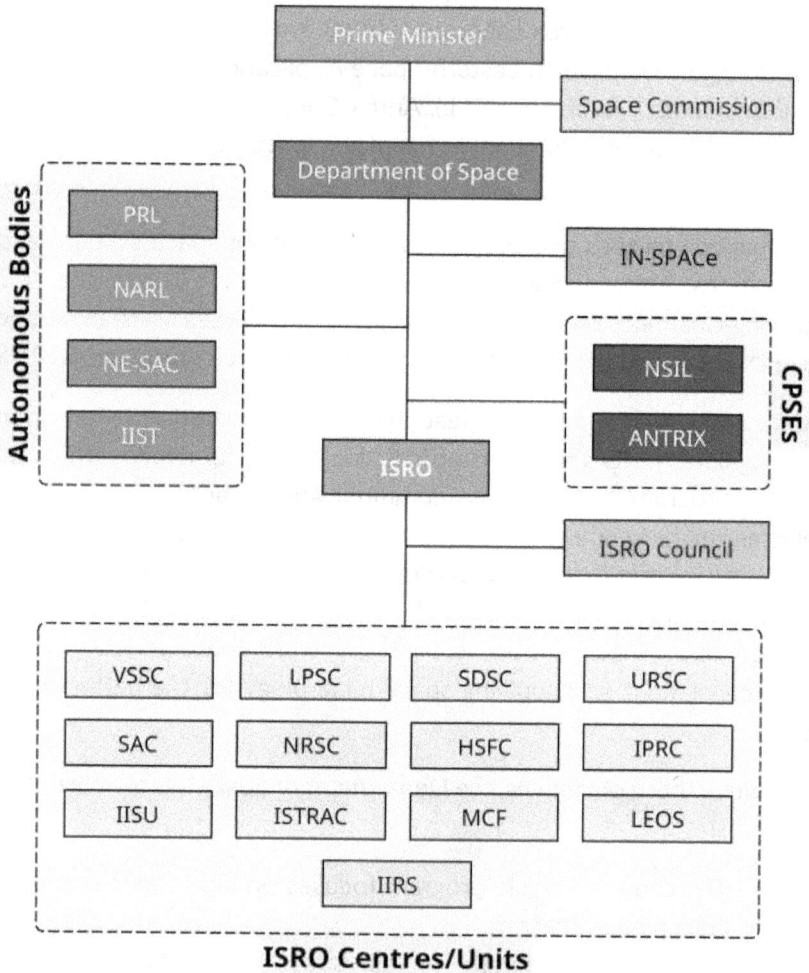

```
                        Prime Minister
                              |
                              |------------ Space Commission
                              |
                     Department of Space
                              |
                              |------------ IN-SPACe
   Autonomous Bodies          |
   ┌─────────────┐            |
   │    PRL      │            |
   │   NARL      │            |------------ NSIL / ANTRIX (CPSEs)
   │  NE-SAC     │            |
   │   IIST      │          ISRO
   └─────────────┘            |
                              |------------ ISRO Council
                              |
```

VSSC	LPSC	SDSC	URSC
SAC	NRSC	HSFC	IPRC
IISU	ISTRAC	MCF	LEOS
	IIRS		

ISRO Centres/Units

Within DoS, the central entity is the Indian Space Research Organisation (ISRO), India's national space agency. ISRO operates under the Department of Space, under the purview of the Prime Minister. It encompasses space applications, exploration, international collaboration, and technological advancements in the space sector. Based in Bangalore, ISRO coordinates various agencies and institutes dedicated to space research and technology advancement, each playing a crucial role in achieving ISRO's objectives. With

13 significant research facilities, ISRO has the means to make significant strides in space technology and exploration.

Lets have a look at the various agencies and institutes that work with ISRO.

Satish Dhawan Space Centre:

In the early 1960s, the Indian Space Research Organisation (ISRO) was established to address various functional requirements related to space exploration. Spread across different centres across India, one of the notable centres is the Satish Dhawan Space Centre (SDSC SHAR), located on the spindle-shaped island of Sriharikota in the Bay of Bengal. Initially, a tiny fishing village inhabited by tribal people, Sriharikota underwent a remarkable transformation over the years to become a world-class launch base strategically located away from major cities.

In 1969, the sounding rocket complex was established at SDSC SHAR for atmospheric experiments using Rohini sounding rockets, which continues to be operational today. The centre houses colossal facilities and plans for solid propellant manufacturing, including the first solid propellant space booster plant called SPROB. This plant played a crucial role in processing solid motor segments for various launch vehicles, including SLV3, ASLV, PSLV, and GSLV. Over time, SDSC SHAR developed a dedicated solid propellant plant called SPP to meet the diverse demands of the launch vehicle program.

SDSC SHAR conceived and realized the first launch pad for the Polar Satellite Launch Vehicle (PSLV) to enhance the launch capabilities and reduce turnaround time between launches. Operating on the concept of "integrate on the pad," the launch pad allowed for extensive checkout systems and liquid propellant servicing facilities. Subsequently, to accommodate heavier rockets like GSLV Mk3, SDSC SHAR migrated to the "integrate transfer launch" concept and added a low bay solid stage assembly building.

In 1971, SDSC SHAR became operational when it launched an RH-125-sounding rocket. The centre attempted its first orbital satellite launch, Rohini 1A, aboard a Satellite Launch Vehicle on 10 August 1979. Despite a failure in thrust vectoring during the rocket's second stage, the centre continued its efforts to achieve success in subsequent missions. The achievements of SDSC SHAR continued to grow over the years. On 5 September 2002, it was renamed the "Satish Dhawan Space Centre SHAR" in memory of Satish Dhawan, the former chairman of ISRO. The centre now boasts two launch pads, with the second one built in 2005. This universal launch pad accommodates all of ISRO's launch vehicles and allows multiple launches in a year.

The significance of SDSC SHAR was further highlighted with the successful launch of India's lunar orbiter Chandrayaan-1 on 22 October 2008 and the Mars orbiter Mangalyaan on 5 November 2013, which was successfully placed into Mars orbit on 24 September 2014. Under the Indian Human Spaceflight Programme, SDSC SHAR is gearing up to augment its existing launch facilities to meet the target of launching a crewed spacecraft called Gaganyaan. Currently led by Arumugam Rajarajan, who assumed office in July 2019, the centre continues to be a pivotal part of India's space exploration journey, contributing to advancements in space science and technology.

Vikram Sarabhai Space Centre (VSSC)

VSSC is located in Thiruvananthapuram and was formed on 21 November 1963. It holds the top position among ISRO facilities and specializes in developing launch vehicle technology. VSSC is responsible for designing and perfecting this technology. The centre actively engages in research across various fields like aeronautics, avionics, materials, and more. It is crucial in designing, manufacturing, analyzing, and testing subsystems for different missions. This is supported by program planning, technology transfer, industry coordination, and safety assurance.

VSSC's extensions are Valiamala, which focuses on mechanisms and testing, and Vattiyoorkavu, which is dedicated to composite development. Additionally, VSSC set up the Ammonium Perchlorate Experimental Plant (APEP) near Kochi in Aluva. At VSSC, major programs include the Polar Satellite Launch Vehicle (PSLV), Geosynchronous Satellite Launch Vehicle (GSLV), Rohini Sounding Rockets, as well as the development of GSLV Mk III, Reusable Launch Vehicles, advanced technology vehicles, air-breathing propulsion, and critical technologies for human spaceflight.

Liquid Propulsion Systems Centre (LPSC)

The Liquid Propulsion Systems Centre (LPSC) is an ISRO research and development centre. It operates through two units in Valiamala, Thiruvananthapuram (Kerala), and Bengaluru (Karnataka). The ISRO Propulsion Complex in Mahendragiri, Tamil Nadu, further supports it.

At the Thiruvananthapuram unit, LPSC focuses on liquid and cryogenic propulsion stages for launch vehicles, plus control systems for both launch vehicles and spacecraft. In Bengaluru, precision fabrication, transducer development, and satellite propulsion system integration happen. The ISRO Propulsion Complex in Mahendragiri handles developmental, flight, assembly, and integration tests. LPSC develops liquid propellant stages for PSLV, control systems for SLV-3, ASLV, PSLV, GSLV, and satellite propulsion systems like INSAT and IRS. They also make pressure transducers. Their significant achievement is the indigenous cryogenic upper stage for GSLV, successfully tested by ISRO in August 2007.

Valiamala's LPSC handles R&D, system design, engineering, and project management. It includes entities for fluid control components, materials and manufacturing, earth storable and cryogenic propulsion. LPSC Bengaluru designs propulsion systems for communication satellites' scientific missions and develops transducers and sensors.

U R Rao Satellite Centre (URSC)

The U R Rao Satellite Centre (URSC), formerly known as ISRO Satellite Centre (ISAC), was formed on May 11, 1972. Initially established in Bangalore's Peenya Industrial Estates as the Indian Scientific Satellite Project (ISSP), it was later renamed to honour Dr. Udupi Ramachandra Rao, ISRO's former Chairman and ISAC's founding director, becoming the URSC on April 2, 2018. In 2018, the centre achieved a milestone by launching its 100th satellite.

URSC is the primary hub for creating satellites and advancing satellite technologies. These satellites serve various purposes, including Communication, Navigation, Meteorology, Remote Sensing, Space Science, and interplanetary exploration. The centre's focus also extends to future mission technologies. URSC boasts cutting-edge facilities for end-to-end satellite building. The ISRO Satellite Integration and Test Establishment (ISITE) features advanced clean rooms, a 6.5-meter thermo vacuum chamber, a 29-ton vibration facility, a Compact Antenna Test Facility, and an acoustic test facility. Communication and Navigation Spacecraft assembly, integration, and testing take place here. Additionally, a dedicated facility for producing standardized subsystems is set up at ISITE.

URSC's Laboratory for Electro-Optics System (LEOS) in Peenya, Bengaluru, researches, develops, and produces sensors for ISRO programs. The centre collaborates with private and public sector industries to realize standardized satellite hardware.

Space Applications Centre (SAC)

The Space Applications Centre (SAC), based in Ahmedabad, was established on October 25, 1972, under the guidance of Dr. Vikram Sarabhai. It spans two campuses and conducts a range of multidisciplinary activities. The Centre specializes in developing instruments and payloads for space and airborne platforms, catering to national development and societal welfare. These applications play a pivotal role in the country's communication, navigation, and remote sensing.

SAC has played significant roles in scientific and planetary missions like Chandrayaan-1 and Mars Orbiter Mission. It's responsible for crafting communication transponders used by the government and private sector for VSAT, DTH, Internet, broadcasting, and telephones. Apart from these achievements, SAC designs and develops optical and microwave sensors for satellites, signal and image processing software, and GIS software—their applications span Geosciences, Agriculture, Environment, Climate Change, Oceanography, Atmosphere, and more. The Centre boasts cutting-edge facilities, including payload integration labs, fabrication facilities, and environmental test centres.

SAC's collaborations with industry, academia, and national and international institutes fuel research and development efforts. The Centre also hosts nine-month postgraduate diploma courses in satellite meteorology and communication under the Center for Space Science and Technology Education (CSSTE-AP). One major accomplishment is the free space quantum communication technology demonstration in March 2021 through the Quantum Experiments using Satellite Technology (QuEST) project. This breakthrough involved successfully establishing quantum communication between two ISRO ground stations. SAC's work extends to payloads for INSAT and IRS satellites and remains a hub for advancing space technology and education.

National Remote Sensing Centre (NRSC)

Established in September 1974, the National Remote Sensing Centre (NRSC) is located in Hyderabad, India. NRSC's primary role is acquiring and processing remote sensing satellite data, disseminating this data and offering support for disaster management. NRSC operates a data reception station near Hyderabad in Shadnagar to facilitate this, capturing information from Indian and foreign remote sensing satellites.

NRSC's ground station in Shadnagar gathers Earth Observation data from various satellites, contributing to applications and projects in collaboration with users. The Aerial Services and Digital Mapping (ASDM) Area within NRSC provides comprehensive aerial remote sensing services, aiding applications like aerial photography, digital mapping, infrastructure planning, and more.

Regional Remote Sensing Centres (RRSCs) function across different regions, supporting specific remote sensing tasks regionally and nationally. These centres engage in application projects that span natural resources. RRSCs are also involved in software development, customization, and packaging to meet user needs. They conduct training programs to educate users in geospatial technology, focusing on digital image processing and Geographical Information System (GIS) applications. NRSC plays a critical role in remote sensing and supporting a range of applications, contributing significantly to India's space capabilities.

Human Space Flight Centre (HSFC)

ISRO established the Human Space Flight Centre (HSFC) on January 30, 2019, with its headquarters in Bangalore, Karnataka, India. The primary purpose of HSFC is to oversee the execution of the Gaganyaan project, which involves sending humans into space. The journey towards crewed space missions commenced in 2007 with the Space Capsule Recovery Experiment (SRE), a 600 kg capsule launched using the Polar Satellite Launch Vehicle (PSLV) rocket. The capsule was successfully retrieved 12 days later, marking a crucial step in ISRO's human spaceflight endeavours.

HSFC holds responsibility for the entire Gaganyaan mission, encompassing mission planning, engineering systems for crew survival in space, crew selection and training, and continued efforts for sustained human spaceflight missions. It collaborates with existing ISRO centres to implement the development flight of Gaganyaan as part of the Human Space Flight Programme. Dr S Unnikrishnan Nair serves as the founder and Director of HSFC, guiding the organization towards achieving milestones in India's human spaceflight journey. Through HSFC's endeavours, ISRO aims to make significant strides in human space exploration and contribute to the global space community.

ISRO Propulsion Complex (IPRC)

The ISRO Propulsion Complex (IPRC) in Mahendragiri is vital in advancing cutting-edge propulsion technology for India's space program. Formerly known as LPSC, Mahendragiri, it was renamed IPRC on February 01, 2014, to reflect its expanding role in our nation's space journey.

IPRC is equipped with advanced facilities that enable activities like assembling, integrating, and testing earth-storable propellant engines, cryogenic engines, and stages for launch vehicles. It conducts high-altitude tests for upper-stage engines, spacecraft thrusters, and their subsystems. Additionally, IPRC produces and supplies cryogenic propellants, a crucial component of India's cryogenic rocket program. A Semi-cryogenic Cold Flow Test facility (SCFT) is established at IPRC to develop and test semi-cryogenic engine subsystems.

Furthermore, IPRC provides Storable Liquid Propellants for ISRO's launch vehicles and satellite projects. The complex is dedicated to ensuring top-notch quality, adhering to the zero-defect requirement of ISRO's space program, and maintaining high levels of safety and reliability.

In addition to its operational role, IPRC engages in Research & Development (R&D) and Technology Development Programs (TDP) to continuously enhance its contributions to India's space program. Through its endeavours, IPRC remains a pivotal hub in advancing propulsion technology, driving India's space capabilities forward.

ISRO Inertial Systems Unit (IISU)

The Indian Space Research Organisation (ISRO) established the ISRO Inertial Systems Unit (IISU) on August 15, 1969, in Vattiyoorkavu, Thiruvananthapuram. It focuses on designing and developing Inertial Systems for ISRO's Launch Vehicles and Spacecraft programs. They create essential systems like Inertial Navigation Systems based on mechanical and optical gyros, Attitude Reference Systems, Rate Gyro Packages, and Accelerometer Packages. These systems are indigenously developed and are utilized in various ISRO missions.

Additionally, IISU is responsible for crafting Actuators and Mechanisms for spacecraft and related applications. It is actively involved in ongoing Research and Development efforts. The unit utilizes its accumulated experience and knowledge to refine existing sensors and systems while pioneering new technologies. IISU has launched advanced technology development programs, positioning itself as a Center of Excellence in Inertial Sensors and Systems.

Through its work, IISU aims to ensure that these systems are efficient, cost-effective, reliable, and aligned with global trends. As a significant contributor to ISRO's capabilities, IISU is crucial in advancing India's space technology.

ISRO Telemetry, Tracking and Command Network (ISTRAC)

The ISRO Telemetry, Tracking and Command Network (ISTRAC) in Bengaluru is crucial in the success of ISRO's satellite and launch vehicle missions. Its primary purpose is to provide tracking support throughout all stages of ISRO's endeavours.

ISTRAC's key objectives encompass several critical areas. It oversees the mission operations of operational remote sensing and scientific satellites. It ensures Telemetry, Tracking, and Command (TTC) services from the launch vehicle's liftoff until the satellite is positioned in its orbit, estimating its initial orbital parameters. The centre engages in hardware and software development to enhance its capabilities in delivering flawless TTC and Mission Operations services.

To achieve these goals, ISTRAC has set up a network of ground stations across various locations, including Bengaluru, Lucknow, Mauritius, Sriharikota, Port Blair, Thiruvananthapuram, Brunei, Biak (Indonesia), and the Deep Space Network Stations.

Besides its established TTC role, ISTRAC is tasked with supporting Deep Space Missions of ISRO. It works on developing radar systems for tracking launch vehicles and meteorological applications. The centre takes on responsibilities like establishing and operating the ground segment for the Indian Regional Navigational Satellite System, providing services for Search and rescue and Disaster Management, and supporting space-based services such as telemedicine, Village Resource Centre (VRC), and tele-education.

Master Control Facility (MCF)

The Master Control Facility (MCF) is a vital establishment of ISRO located in Hassan, Karnataka, India. Founded in 1982, its primary purpose is to oversee and manage geostationary and geosynchronous satellites launched by ISRO. Initially, it was the sole Master Control Facility until another was established in Bhopal in 2005.

MCF is divided into three internal sections: the Spacecraft Control Centre, the Mission Control Centre, and the Earth Station. The Spacecraft Control Centre commands satellites, transmitting instructions to carry out specific operations. The Mission Control Centre leads during a satellite's launch and its initial time in space. Here, the Mission Director and module designers monitor the satellite's performance by analyzing telemetry signals. Any deviation from the set parameters triggers an automatic alarm, leading designers to collaborate with the Mission Director and adjust satellite functions accordingly.

The Earth station employs an array of antennas to establish communication between the MCF and the satellite. The connection from the earth station to the satellite is the UPLINK, while the link from the satellite to the earth station is the DOWNLINK. MCF features three full-motion and around a dozen limited-motion antennas. These antennas facilitate sending commands to the satellite and receiving signals, ensuring the satellite maintains its orbit.

Laboratory for Electro-Optics Systems (LEOS)

In 1993, the Laboratory for Electro-Optics Systems (LEOS) was established in Bengaluru, where India's first satellite, Aryabhatta, was crafted in 1975. LEOS, a crucial unit of ISRO, operates from the Peenya Industrial Estate and specializes in designing, developing, and producing Attitude Sensors for Low Earth Orbit (LEO), Geostationary Orbit (GEO), and interplanetary missions. It also creates and delivers Optical Systems for remote sensing and meteorological payloads.

LEOS boasts state-of-the-art fabrication, testing, and coating facilities. The lab delves into cutting-edge technologies like 3-axis Fiber Optics Gyro, Optical Communication, MEMS, Nanotechnology, Detectors, and developing Science Payloads for upcoming space missions. Over the years, LEOS has developed sensors for tracking Earth and Stars, contributing to satellites like Aryabhatta, Bhaskara, Apple, IRS, SROSS, and INSAT-2. The lab actively participated in India's inaugural Moon mission, Chandrayaan-1, and will contribute to ISRO's upcoming Sun mission, Aditya-L1. Through its innovations, LEOS significantly contributes to India's space journey.

Indian Institute of Remote Sensing (IIRS)

Formerly known as the Indian Photo-interpretation Institute (IPI), this institute was established on 21 April 1966 under the Survey of India's (SOI) guidance. The inspiration for its creation came from India's first prime minister, Pandit Jawahar Lal Nehru, during his visit to the Netherlands in 1957. The institute's building in Dehradun, Uttarakhand, was inaugurated on 27 May 1972.

IIRS is a prominent institution focused on enhancing Remote Sensing and Geo-informatics expertise and their applications through postgraduate-level education and training programs. The institute also supports the Centre for Space Science and Technology Education in Asia and the Pacific (CSSTE-AP), affiliated with the United Nations.

IIRS designs its training and education programs to cater to a wide range of individuals, including professionals at various levels, fresh graduates, researchers, academics, and decision-makers. The courses span from one week to two years in duration. As of 30 April 2011, IIRS became a distinct entity within ISRO. The institute emphasises Training, Education, and Research while focusing on advancing expertise in Microwave Remote Sensing, Hyperspectral Remote Sensing, and Climate studies. Through its endeavours, IIRS contributes significantly to enhancing remote sensing and geoinformatics capabilities.

In addition to the agencies and institutes mentioned earlier, the Department of Space also supervises the operations of the following four autonomous bodies. Here is a breakdown of their work and areas of research.

Physical Research Laboratory (PRL)

The Physical Research Laboratory (PRL), an autonomous unit of the Department of Space, serves as a National Research Institute for space and Allied Sciences. Founded on 11 November 1947 by Dr Vikram Sarabhai at Ahmedabad, its initial focus was researching cosmic rays and the upper atmosphere due to a limited understanding of high-energy particles bombarding Earth.

PRL conducts cutting-edge research in several fields, including Astronomy and Astrophysics, Solar Physics, Planetary Science and Exploration, Space and Atmospheric Sciences, Geosciences, Theoretical Physics, Atomic, Molecular, and Optical Physics, and Astrochemistry. PRL plays a significant role in planetary exploration, achieving notable progress in planetary sciences and exploration. It engages in stellar and solar astronomy studies at the Infra-red Observatory in Mt. Abu and a Solar Observatory in Udaipur. The Thaltej, Ahmedabad campus hosts the Planetary Exploration (PLANEX) program.

The research divisions within PRL are as follows:

Astronomy and Astrophysics:

This division delves into optical, infrared, X-ray, and radio wavelengths to study galactic and extragalactic cosmic phenomena.

Atomic, Molecular, and Optical Physics:

With an interdisciplinary approach, this division studies topics like astrochemistry, quantum mechanics' foundations, and luminescence dating. It also investigates the classical and quantum properties of light, atoms, molecules, and molecular clusters across the electromagnetic spectrum and high-energy electrons.

Planetary Sciences:

This division aims to characterize processes in the early Solar System, focusing on the origin and evolution of the solar system, particularly inner planets. It comprehends planet atmospheres' physical and chemical processes through simulations, observations, and modelling. The Planetary Science and Exploration Program (PLANEX) is also part of this division.

Theoretical Physics:

Engaging in theoretical and phenomenological research, this division explores atomic physics, condensed matter physics, gravitation, astroparticle physics, non-equilibrium phenomena, and particle physics.

Space and Atmospheric Sciences:

This division studies radiative, chemical, ionization, and dynamical processes in Earth's atmosphere through various methods, including in-situ rocket and balloon experiments, laboratory experiments, and theoretical simulations.

Geosciences:

This department examines Earth's origin and evolution and its components by focusing on geochronology, geochemistry, glaciology, oceanography, and palaeoclimatology.

National Atmospheric Research Laboratory(NARL)

The National Atmospheric Research Laboratory (NARL) is a self-governing research institute funded by the Department of Space (DOS). Established in 1992 as the National Mesosphere-Stratosphere-Troposphere (MST) Radar Facility (NMRF), it engages in fundamental and applied research in Atmospheric Sciences. Over time, it expanded its facilities to include Mie/Rayleigh Lidar, wind profiler, rain gauge, disdrometer, and weather stations. In 2005, NMRF transformed into NARL, a comprehensive atmospheric research institute.

Situated at Gadanki near Tirupati, NARL is an atmospheric research hub dedicated to technology development, observations, data management, assimilation, and modelling. NARL's research comprises seven core groups: Radar Application and Development, Ionospheric and Space Research, Atmospheric Structure and Dynamics, Cloud and Convective Systems, Aerosols, Radiation and Trace Gases, Weather and Climate Research, and Computers and Data Management. The institute also undertakes specific projects, including LiDAR and Advanced Space-borne Instrument Development.

Supported by DOS, NARL plays a vital role in advancing our understanding of Earth's atmosphere. Its diverse research groups collaborate to explore various atmospheric aspects, from ionospheric behaviour to cloud systems and climate patterns. NARL's comprehensive approach, encompassing technology, observations, and modelling, contributes significantly to studying atmospheric sciences in India and beyond.

North Eastern-Space Applications Centre (NE-SAC)

The North Eastern-Space Applications Centre (NE-SAC) is an autonomous research institute funded by the Department of Space (DOS) in collaboration with the North Eastern Council (NEC). Established on 5 September 2000, NE-SAC is situated in Shillong with a specific mission to leverage space science and technology for the overall development of India's North Eastern Region (NER).

NE-SAC emerged as a joint effort of DOS and NEC to harness the potential of remote sensing technology for exploring natural resources in the North-Eastern states. It aims to accelerate the growth of these states by applying space science and technology to various sectors. The centre's activities encompass the development of advanced technological infrastructure to drive the comprehensive development of the NER.

NE-SAC is pivotal for implementing regional and national programs related to resource management, infrastructure planning, healthcare, education, disaster management, atmospheric science research, and more. It collaborates with State Remote Sensing Application Centres in the NER to execute these initiatives effectively. The centre's portfolio includes projects sponsored by regional user agencies and research and development endeavours under critical programs like Earth Observation Applications Mission, ISRO Geo-sphere Biosphere Programme, Satellite Communications, Disaster Management Support, and Space Science Programs.

Indian Institute of Space Science and Technology (IIST)

The Indian Institute of Space Science and Technology (IIST) is an autonomous research institute funded by the Department of Space. Situated in Valiamala, Nedumangad, Thiruvananthapuram, Kerala, IIST is a unique government-aided institute and deemed university solely dedicated to the study and research of space science. It holds the distinction of being the first university in Asia with such a specialized focus.

Established on 14 September 2007 and inaugurated by G. Madhavan Nair, the then Chairman of ISRO, IIST operates as Asia's pioneering Space University. Notably, it was graced by the Chancellorship of A. P. J. Abdul Kalam, the former President of India. The institute is designed to address the educational needs of the Indian Space Programme, offering a range of educational programs, including undergraduate, postgraduate, doctoral, and post-doctoral courses in space science, technology, and applications.

With an unwavering commitment to excellence, IIST emphasizes top-notch education, learning, and research. The institute serves as a hub for cutting-edge research and development in space studies, contributing substantially to the growth of the Indian Space Programme. IIST functions as a think-tank, constantly exploring innovative avenues for space exploration.

The institute boasts a robust faculty team of nearly 100 members across seven departments. Located approximately 20 km from Thiruvananthapuram city, IIST operates as a residential institution in Valiamala. Since its inception, IIST has significantly contributed to advancing space science, nurturing bright minds, and facilitating pioneering research endeavours, cementing its role as a premier institution in space education and exploration.

Antrix Corporation Limited

Antrix Corporation Limited, based in Bengaluru, is a government-owned company that operates under the Department of Space. Established on September 28, 1992, Antrix serves as the commercial and marketing arm of ISRO. Its primary role involves promoting, commercially delivering, and marketing the various products and services developed by ISRO.

Antrix Corporation Limited (ACL) is a wholly owned Government of India Company. Initially established as a private limited company, ACL's primary purpose is to market and promote space products, provide technical consultancy services, and transfer technologies developed by ISRO. Additionally, it aims to foster the growth of space-related industrial capabilities in India.

As the commercial and marketing branch of ISRO, Antrix offers space products and services to international customers globally. Equipped with advanced facilities, Antrix provides comprehensive solutions for a wide range of space products. These offerings include hardware, software, subsystems, and even complex spacecraft for applications like communications, earth observation, and scientific missions. Antrix offers remote sensing data, transponder leasing, launch services using operational vehicles like PSLV and GSLV, mission support, and consultancy and training services.

Over the years, Antrix has collaborated with various international entities, including EADS Astrium, Intelsat, Avanti Group, WorldSpace, Inmarsat, SES World Skies, Measat, Singtel, and other space institutions in Europe, the Middle East, and Southeast Asia. Through such collaborations, Antrix has played a vital role in promoting and expanding ISRO's space products and services on a global scale.

NewSpace India Limited (NSIL)

The NewSpace India Limited (NSIL) is a Government of India Public Sector Undertaking (PSU) and the commercial branch of the Indian Space Research Organisation (ISRO). Established on March 6, 2019, NSIL is responsible for producing, assembling, and integrating launch vehicles in collaboration with industry consortiums.

The core objective of NSIL is to enhance private sector involvement in Indian space programs. NSIL was created with specific goals, including the transfer of small satellite technology to the industry. It acts as an intermediary by obtaining licenses from the Department of Space (DoS)/ISRO and sub-licensing them to the industry.

Additionally, NSIL aims to collaborate with the private sector to manufacture Small Satellite Launch Vehicles (SSLV) and contribute to producing Polar Satellite Launch Vehicles (PSLV) through Indian industries. It also takes on the role of producing and marketing space-based products and services, encompassing both launch services and applications. NSIL is entrusted with transferring technologies developed by various ISRO Centres and units under the Department of Space. It facilitates the domestic marketing of spin-off technologies and products and extends its outreach to international markets.

Indian National Space Promotion and Authorization Center (IN-SPACe)

The Indian National Space Promotion and Authorization Centre (IN–SPACe) operates as an autonomous agency under the Government of India's Department of Space. The Indian space program, spanning over five decades, has concentrated on practical applications and accessibility to the general populace. This commitment has propelled it to be counted among the world's top six space agencies. The Indian Space Research Organisation (ISRO) boasts an extensive fleet of GEO communication and LEO remote sensing satellites, addressing the escalating needs for swift communication and reliable earth observation.

In June 2020, a historic decision led by Prime Minister Narendra Modi was made to open up the nation's space sector. This initiative aimed to integrate the Indian private industry into various space activities. As a result, the Indian National Space Promotion and Authorisation Centre (IN-SPACe) was established as an independent single-window agency within the Department of Space (DOS). Designed to oversee private entities' space sector activities, IN-SPACe plays a pivotal role in boosting India's private space sector economy.

Over the years, ISRO has cultivated numerous industries and Micro, Small, and Medium Enterprises (MSMEs) as partners in creating launch vehicles and satellites. Recognising the untapped potential within the country and the global growth of the space sector, the decision to enable NewSpace-enabled companies (NGEs) to undertake independent space activities was a strategic move.

IN-SPACe is responsible for promoting, enabling, authorising, and supervising various space endeavours by NGEs. This encompasses tasks such as constructing launch vehicles, satellites, and space-based services. Furthermore, IN-SPACe facilitates sharing space infrastructure and premises controlled by DOS/ISRO, along with establishing new space infrastructure and facilities.

Three Directorates carry out the agency's operations: Promotion Directorate (PD), Technical Directorate (TD), and Program Management and Authorization Directorate (PMAD). IN-SPACe's creation and mandate reflect India's strategic vision of leveraging its resources, technical capabilities, and burgeoning industries to foster a thriving private space sector and cement its place as a formidable player in space exploration.

This book covers only some other centres and institutes; however, their contribution to ISRO's success is equally important. You may find the information about the other centres in the map below.

NavIC

The Indian Space Research Organisation (ISRO) initiated the Indian Regional Navigation Satellite System (IRNSS), colloquially known as NavIC (Navigation with Indian Constellation), to establish an indigenous satellite-based navigation network.

Prompted by the United States' refusal to grant access to the Global Positioning System (GPS) data during the Kargil Conflict in 1999, which left Indian military forces disadvantaged in the Himalayan region, India recognized the necessity of an autonomous navigation system. The Indian government sanctioned the project in May 2006, motivated by the strategic significance of self-reliance in navigation technology.

ISRO marked a significant milestone on May 28, 2013, with the inauguration of a dedicated facility within their Deep Space Network (DSN) as an integral part of the NavIC project. The network encompassed 21 ranging stations strategically positioned across the nation. These stations facilitated the precise orbital determination of satellites and real-time monitoring of navigation signals. Notably, while three satellites occupied the geostationary orbit over the Indian Ocean, satellites at lower altitudes enabled efficient coverage for low-inclination orbits. The NavIC constellation consisted of 8 satellites: IRNSS-1A, IRNSS-1B, IRNSS-1C, IRNSS-1D, IRNSS-1E, IRNSS-1F, IRNSS-1G, and IRNSS-1I. The first satellite was successfully launched on July 1, 2013.

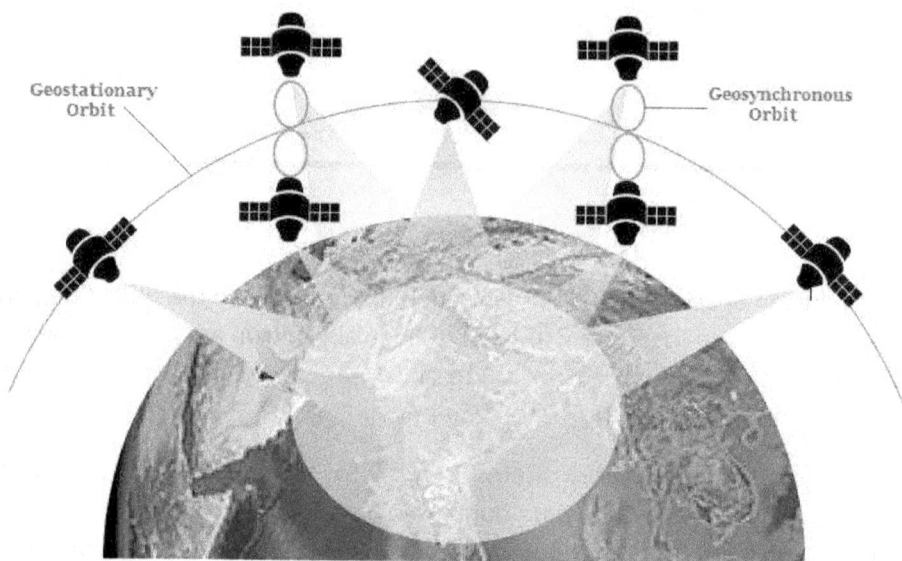

Geostatationary Orbit

Geosynchronous Orbit

NavIC operates as an autonomous regional satellite navigation system, offering accurate real-time positioning and precise timing services. Its coverage spans India and extends up to 1,500 kilometres around its borders. Designed with meticulous precision, the system delivers an impressive absolute position accuracy of less than 10 meters across the Indian landmass and less than 20 meters within the Indian Ocean region.

Looking ahead, ISRO has ambitious plans for NavIC's expansion. The organization intends to launch a series of advanced five-generation satellites equipped with enhanced payloads designed for a lifespan exceeding 12 years. These upcoming satellites, named NVS-01, NVS-02, NVS-03, NVS-04, and NVS-05, will further fortify the existing satellite constellation, bolstering the capabilities and reach of the NavIC network.

The NavIC project is a testament to India's commitment to technological sovereignty and self-sufficiency. It showcases ISRO's technical prowess and underscores the nation's determination to secure its strategic interests in navigation technology, ensuring its military and civilian applications are unhindered by external constraints.

Polar Satellite Launch Vehicle (PSLV)

In response to the need for a more capable launch vehicle, ISRO initiated the development of the Polar Satellite Launch Vehicle in 1978. The primary objective was to create a vehicle capable of delivering a 600 kg payload to a 550 km sun-synchronous orbit from the SHAR (Sriharikota Range) spaceport.

The maiden flight of the PSLV took place on 20 September 1993. The first and second stages of the launch proceeded as expected. However, an unexpected attitude control problem during separation resulted in a collision between the second and third stages, causing the payload to fail to reach its intended orbit.

Despite this initial setback, the Polar Satellite Launch Vehicle eventually achieved success with its second mission in 1994. It demonstrated its capabilities and potential by launching India's Indian Remote Sensing (IRS) satellites into sun-synchronous orbits, a feat previously only offered commercially by Russia. Additionally, the PSLV displayed its versatility by enabling the launch of small-sized satellites into Geostationary Transfer Orbit (GTO).

Over the years, the PSLV has undergone several improvements, enhancing its thrust, efficiency, and weight management in each subsequent version. As a result of these advancements, the PSLV achieved an impressive record, completing 34 launches by November 2014 with no further failures.

The Polar Satellite Launch Vehicle has emerged as a reliable and prominent launch vehicle, especially for low Earth orbit (LEO) satellite missions. It has facilitated numerous significant satellite deployments, including India's ground-breaking missions, such as the first lunar probe, Chandrayaan-1, and its first interplanetary venture, the Mars Orbiter Mission (Mangalyaan). Furthermore, the PSLV played a vital role in deploying India's inaugural space observatory, Astrosat.

One of the key strengths of the PSLV is its reputation as a leading provider of rideshare services for small satellites. The vehicle has been instrumental in numerous multi-satellite deployment campaigns, often accommodating auxiliary payloads alongside the primary Indian payload. As of June 2022, the PSLV successfully launched 345 foreign satellites from 36 countries. One of the most notable achievements in this regard was the PSLV-C37 mission on 15 February 2017, which impressively deployed 104 satellites into sun-synchronous orbit.

The Polar Satellite Launch Vehicle (PSLV) is a testament to ISRO's determination and commitment to advancing India's space exploration capabilities. Despite early challenges, the PSLV has proven its reliability and versatility, earning its place as a significant player in the global space community. The vehicle's remarkable track record and achievements underscore its vital role in supporting Indian and international satellite launches, positioning India as a prominent player in space technology and exploration. The success of the PSLV has further bolstered ISRO's reputation as a reliable and cost-effective space agency.

With its consistent performance and diverse range of capabilities, the PSLV has become the workhorse of ISRO's satellite launch missions. It has served India's needs for communication, Earth observation, and scientific research and has been a preferred choice for foreign countries seeking cost-effective satellite launch solutions.

As ISRO continues to push the boundaries of space exploration, the PSLV remains a crucial asset in its quest for further advancements. The development of more powerful and technologically advanced launch vehicles, such as the Geosynchronous Satellite Launch Vehicle (GSLV) and the Geosynchronous Satellite Launch Vehicle Mark III (GSLV Mk III), complements the PSLV's capabilities and ensures ISRO's position as a significant player in the global space arena.

PSLV Variants:

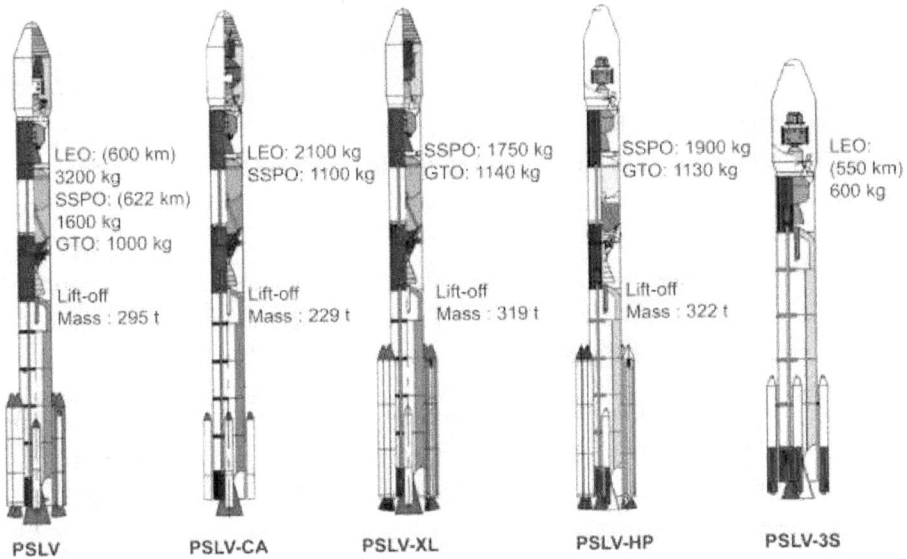

| PSLV | PSLV-CA | PSLV-XL | PSLV-HP | PSLV-3S |

ISRO's PSLV has been developed in various configurations to meet diverse mission requirements. These variants are designed to accommodate different payload capacities, ranging from 600 kg in low earth orbit (LEO) to 1,900 kilograms in a sun-synchronous orbit (SSO). Currently, two operational versions of the PSLV are in use, each offering distinct capabilities.

1. PSLV-G (Generic Version): The PSLV-G, also known as the standard version, has four stages that alternate between solid and liquid propulsion systems. It features six strap-on boosters (PSOM or S9), each with a 9-tonne propellant load. With this configuration, the PSLV-G can launch payloads weighing up to 1,678 kg into a sun-synchronous orbit at 622 km altitude.

2. PSLV-CA (Core Alone): The PSLV-CA model, which debuted on 23 April 2007, differs from the standard variant by excluding the six strap-on boosters. However, two SITVC tanks with Roll Control Thruster modules and two cylindrical aerodynamic stabilizers remain attached to the first stage. The fourth stage of the PSLV-CA variant has 400 kg less propellant than the standard version. Currently, the PSLV-CA can launch payloads weighing up to 1,100 kilograms into a sun-synchronous orbit at 622 km altitude.

3. PSLV-XL (Extended Version): The PSLV-XL is an upgraded version of the standard PSLV configuration, featuring more powerful and stretched strap-on boosters with a 12-tonne propellant load. Weighing 320 tonnes at lift-off, the PSLV-XL employs larger strap-on motors to achieve a higher payload capability. The successful testing of the improved version of the strap-on booster on 29 December 2005 paved the way for its maiden use in the launch of Chandrayaan-1 by PSLV-C11. The payload capacity of the PSLV-XL variant is 1,800 kg to a sun-synchronous orbit.

4. PSLV-DL (Dual-Strap-On Variant): The PSLV-DL variant utilizes only two strap-on boosters, each carrying a 12-tonne propellant load. PSLV-C44, launched on 24 January 2019, was the first flight to use the PSLV-DL variant of the Polar Satellite Launch Vehicle. This configuration can launch payloads weighing up to 1,257 kg into a 600 km sun-synchronous orbit.

5. PSLV-QL (Quad-Strap-On Variant): The PSLV-QL variant incorporates four ground-lit strap-on boosters, each loaded with 12 tonnes of propellant. Its inaugural flight, PSLV-C45, took place on 1 April 2019. The PSLV-QL variant can launch payloads weighing up to 1,523 kg into a 600 km sun-synchronous orbit.

These various configurations of the PSLV demonstrate ISRO's commitment to adapting and enhancing its launch capabilities to accommodate a wide range of missions. PSLV launched a total of 62 (27 Indian and 35 foreign) satellites, including Resourcesat-2, Youthsat, X-sat, Cartosat-2B, Alsat-2A, Studsat, Oceansat-2, six nanosatellites, Risat-2, Anusat, Cartosat-2A, IMS-1, DLR-Tubsat, Space capsule Recovery Experiment (SRE-1), KALPANA-1 and several others satellites. India also used it as a launch vehicle for the Chandrayaan-1 lunar probe mission. The PSLV's versatility and success in deploying various satellites have established ISRO as a significant player in the global space arena. As India continues to pursue ambitious space exploration endeavours, the PSLV will remain a cornerstone of its space launch capabilities.

Indian Remote Sensing Programme

The Indian Remote Sensing Programme was conceived with the noble vision of harnessing space technologies for the betterment of humanity and national development. It entailed the establishment of three key capabilities to achieve its objectives:

1. The design, construction, and launch of satellites into a Sun-synchronous orbit.

2. The creation and operation of ground stations responsible for spacecraft control, data transfer, data processing, and archival.

3. The utilization of acquired data for various applications on the ground.

A significant milestone in this endeavour was the successful launch of IRS-1A, the maiden satellite of the indigenous state-of-the-art remote sensing satellite series, into a polar Sun-synchronous orbit on 17 March 1988. The achievement was celebrated with the release of a commemorative postage stamp by the Government of India.

The roots of the Indian Remote Sensing Programme can be traced back to the successful demonstration flights of the Bhaskara-1 and Bhaskara-2 satellites in 1979 and 1981, respectively. These missions laid the foundation for developing the IRS program to bolster the national economy through applications in vital sectors such as agriculture, water resources, forestry and ecology, geology, watersheds, marine fisheries, and coastal management.

While the launch of IRS-P1 on 20 September 1993 encountered failure due to a launch issue with the PSLV, subsequent launches were met with success, marking the program's resilience and determination.

India established the National Natural Resources Management System (NNRMS) to support its objectives, with the Department of Space (DOS) serving as the nodal agency, providing operational remote sensing data services. The data acquired from the IRS satellites finds applications across the globe, making a positive impact in various domains. The availability of high-resolution satellites has expanded the range of applications, encompassing urban sprawl, infrastructure planning, and large-scale mapping initiatives.

With an impressive fleet of 11 operational satellites, the IRS system is the world's largest constellation of remote sensing satellites dedicated to civilian use. Positioned in polar Sun-synchronous orbits, these satellites provide invaluable data in diverse spatial, spectral, and temporal resolutions. As a testament to its success and unwavering commitment, the Indian Remote Sensing Programme celebrated its 25th anniversary of successful operations on 17 March 2013.

Oceansat-1

Oceansat-1, also known as IRS-P4, is significant in India's space exploration as it was the country's first satellite primarily dedicated to ocean applications. As part of the Indian Remote Sensing Programme satellite series, Oceansat-1 was equipped with an Ocean Colour Monitor (OCM) and a Multi-frequency Scanning Microwave Radiometer (MSMR), enabling extensive oceanographic studies.

Including OCM and MSMR in Oceansat-1 represented a significant advancement in India's satellite capabilities, enhancing the scope of the IRS satellite system. Before Oceansat-1, the IRS system comprised four satellites: IRS-1B, IRS-1C, IRS-P3, and IRS-1D. With the addition of Oceansat-1, remote sensing applications were extended to newer and diverse areas, broadening the scope of ISRO's scientific pursuits.

On May 26, 1999, Oceansat-1 embarked on its journey into space aboard the Polar Satellite Launch Vehicle (PSLV-C2) from the Satish Dhawan Space Centre. The launch proved a resounding success, marking the third triumphant mission for PSLV. Additionally, Oceansat-1 became the eighth satellite in India's esteemed Indian Remote Sensing Programme (IRS) satellite series.

Originally designed with an anticipated lifespan of five years, Oceansat-1 surpassed expectations, serving for an impressive eleven years before completing its mission on August 8, 2010. Throughout its operational years, Oceansat-1 contributed immensely to oceanographic research and significantly advanced India's prowess in satellite-based remote sensing.

Kalpana-1 Satellite:

Kalpana-1, formerly known as METSAT, holds a significant place in ISRO's history as the first exclusive meteorological satellite built by the organization. It was launched on September 12, 2002, using the Polar Satellite Launch Vehicle (PSLV-C4) and marked the debut of ISRO's satellite series dedicated to meteorological applications.

The satellite was named Kalpana-1 by the Indian Prime Minister Atal Bihari Vajpayee on February 5, 2003, in honour of Dr. Kalpana Chawla—an Indian-born American astronaut who tragically lost her life in the Space Shuttle Columbia disaster on February 1, 2003. This renaming was a tribute from the Indian government to commemorate her contributions to space exploration.

Kalpana-1's design featured a three-axis stabilized platform and relied on solar panels to generate up to 550 watts (0.74 hp) of power. The satellite was equipped with a Very High Resolution scanning Radiometer (VHRR), capable of capturing three-band images (visible, infrared, and thermal infrared) at a resolution of 2 km × 2 km (1.2 mi × 1.2 mi). Additionally, it carried a Data Relay Transponder (DRT) payload that facilitated the transmission of weather data to terrestrial platforms.

Among its missions was data collection of atmospheric layers such as clouds, water vapour, and temperature. Furthermore, Kalpana-1 played a crucial role in establishing the I-1000 bus system, specifically designed to cater to the unique service requirements of meteorological payloads for earth imageries. Notably, the METSAT bus, which served as the foundation for Kalpana-1, was later adapted for the Chandrayaan lunar orbiter mission in 2008, showcasing the versatility and adaptability of ISRO's satellite technology. However, after many years of successful service, Kalpana-1 eventually went out of service in mid-2018.

Spy Satellites:

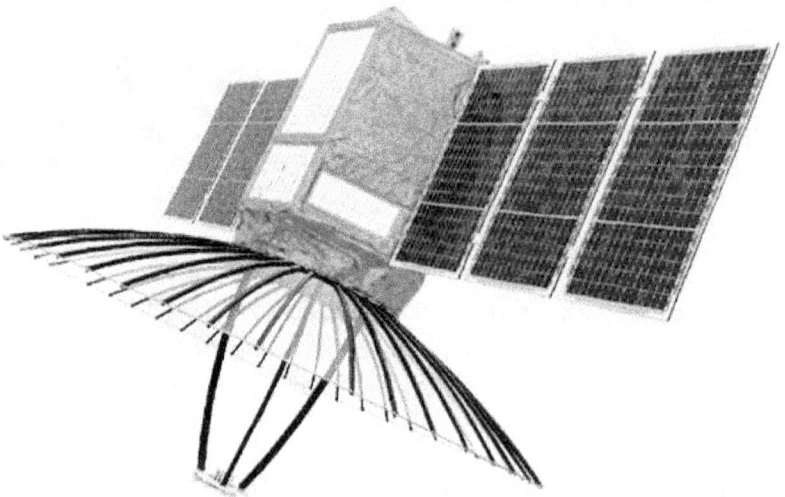

The Technology Experiment Satellite (TES) satellite, weighing 1108 kg, took flight on October 22, 2001, from the Sriharikota High Altitude Range (SHAR). It was a pioneering spacecraft launched by ISRO as a test satellite to demonstrate and validate cutting-edge technologies crucial for future reconnaissance (spy) satellite missions.

TES was equipped with a range of experimental systems, including an attitude and orbit control system, high-torque reaction wheels, a new reaction control system, a light-weight spacecraft structure, a solid-state recorder, an X-band phased array antenna, an improved satellite positioning system, and miniaturized TTC (Telemetry, Tracking, and Command) and power systems. A notable technological innovation of TES was its state-of-the-art two-mirror-on-axis camera optics system.

Incorporating a panchromatic camera in TES allowed for remote sensing experiments, capturing images at an exceptional resolution of one meter. This capability enabled India to become the second country globally, after the United States, with the commercial capacity to provide images of such extraordinary clarity.

TES's remote sensing capabilities found practical applications in various civilian sectors, including mapping, geographical information services, and remote sensing of civilian areas. Moreover, TES was crucial in supporting the US Army during the 11 September 2001 counter-terrorism offensive against the Taliban by providing high-resolution images critical for intelligence and planning operations.

The success of the TES satellite marked a significant milestone for ISRO, as it paved the way for deploying more advanced remote sensing technologies. Subsequently, India saw the emergence of the Integrated Space Cell, jointly operated by the Indian Armed Forces, the Defence Research and Development Organisation (DRDO), and ISRO, serving as the central agency responsible for ensuring the security of the country's space-based military and civilian hardware systems. However, in 2019, this agency was replaced by the Defence Space Agency.

Building upon the accomplishments of the TES mission, India initiated a series of satellite launches with dual purposes. Presently, the country boasts the world's largest constellation of remote sensing satellites, with 14 satellites operational. Among these, at least four satellites have been dedicated to military applications, including GSAT-7, GSAT-6, GSAT-7A, and EMISAT by DRDO. However, it is noteworthy that GSAT-6A launched initially as a dedicated satellite for the army, experienced a communication failure post-launch. Furthermore, as of January 24, 2019, both HySIS and Microsat-R satellites have been designated dual-use satellites, available for military and civilian applications.

India's military satellite program has seen remarkable advancements in deploying notable satellites that significantly contribute to national security and intelligence gathering. Among these critical satellites are:

1) **RISAT-2** - RISAT-2 was built at an accelerated pace following the 2008 Mumbai attacks, launched on April 20, 2009, and was a significant milestone as India's first satellite equipped with a synthetic aperture radar from Israel Aerospace Industries (IAI). Its capabilities included day-night, all-weather monitoring with a remarkable one-meter resolution, enabling potential applications like tracking hostile ships at sea. This Radar imaging satellite is used to monitor India's borders and as part of anti-infiltration and anti-terrorist operations.

2) **CARTOSAT-2** - launched on January 10, 2007, carried a state-of-the-art panchromatic camera capable of capturing black and white pictures with an impressive 80-centimeter spatial resolution. The satellite's agility allowed it to be steered up to 45 degrees along and across its track, providing high-resolution imagery for specific scenes. However, CARTOSAT-2 was decommissioned in 2020.

3) **Cartosat-2A** - launched on April 28, 2008, served as a dedicated satellite for the Indian Armed Forces, boasting capabilities similar to CARTOSAT-2. Its highly agile nature allowed for more frequent imaging of any area.

4) **Cartosat-2B** - launched on July 12, 2010, carried a panchromatic camera with an 80-centimeter resolution, maintaining the high standard of its predecessors. Its agility enabled it to capture multiple spot scene imagery.

5) **GSAT-7** - launched in 2013, was exclusively used by the Indian Navy to monitor the Indian Ocean Region (IOR) and significantly enhance naval operations worldwide through secure, real-time communications.

6) **HySIS (Hyperspectral Imaging Satellite)** - launched on November 29, 2018, played a vital role in providing hyperspectral imaging services, catering to various sectors such as agriculture, forestry, and geography assessment, including coastal zones and inland waterways. The data collected by HySIS was also made accessible to India's defence forces, further augmenting its utility.

7) **GSAT-7A** - launched in December 2018, proved to be an invaluable asset to the Indian Air Force, significantly enhancing network-centric warfare capabilities by effectively interlinking various ground radar stations, airbases, and airborne early warning and control (AWACS) aircraft.

8) **Microsat-R satellite** - dedicated exclusively for the Indian Armed Forces, was successfully launched on January 24, 2019, utilizing the PSLV C-44 rocket. The 760 kg imaging satellite is poised to bolster India's military capabilities further.

9) **EMISAT** - launched on April 1, 2019, represented a significant step forward in reconnaissance and intelligence gathering. As part of DRDO's project Kautilya, EMISAT was designed to provide space-based electronic intelligence (ELINT) crucial for enhancing situational awareness for the Indian Armed Forces.

These remarkable military satellites showcase India's incredible space exploration and technology strides, bolstering its capabilities in remote sensing, communications, intelligence gathering, and overall national security. The successful deployment of these advanced satellites is a testament to ISRO's expertise and growing significance on the global stage.

STUDSAT

In a remarkable display of educational collaboration, the Student Satellite (STUDSAT) emerged as India's pioneering pico-satellite, the brainchild of a consortium comprising seven engineering colleges from Karnataka and Andhra Pradesh. Weighing less than a kilogram, the STUDSAT was conceived to advance space technology within educational institutions and ignite enthusiasm for research and development in miniaturized satellite technology.

Launched triumphantly on July 12, 2010, from the Satish Dhawan Space Centre, the STUDSAT embarked on its journey to orbit the Earth. This momentous achievement marked the onset of a historic communication link between the satellite and ground stations, opening avenues for capturing high-resolution Earth imagery with a remarkable 90-meter resolution.

The inception of this extraordinary project traces back to a quartet of ambitious students from various engineering colleges in Hyderabad and Bangalore who attended the International Astronautical Congress (IAC).

Inspired by this event, they embarked on a journey that led to the creation of the STUDSAT. Elegantly designed as a small rectangular cube with dimensions of 10 cm x 10 cm x 13.5 cm and weighing approximately 950 grams, the STUDSAT ventured into a Sun-synchronous orbit. It etched its place in history by sending its inaugural signal on 12 July 2010, at 11:07 am IST, marking the start of its mission. The STUDSAT didn't just remain a technological marvel; it accomplished its objectives with flying colours. Over time, it secured its place in the Limca Book of Records annals, earning recognition for being the smallest satellite crafted on Indian soil.

As its mission life concluded, the STUDSAT demonstrated its prowess as a remote sensing satellite, capturing Earth's surface imagery at a groundbreaking 90-meter resolution. In doing so, it achieved an unparalleled feat among its "PICO" satellite peers, elevating India's status in the realm of space technology. The STUDSAT saga is one of education, innovation, and the boundless potential that collaboration and passion can unleash. After STUDSAT's success, Russian and Indian students collaborated to develop YouthSat, a joint satellite venture between the Russian Federal Space Agency and ISRO. Utilizing ISRO's Indian Mini Satellite-1 platform, YouthSat and Resourcesat-2 were launched via the Polar Satellite Launch Vehicle on 20 April 2011 from Sriharikota, India.

YouthSat

In a testament to the evolving landscape of international collaboration, following the successful STUDSAT project, Russian students united their efforts with their Indian counterparts to propel the next frontier of satellite development. The outcome was **YouthSat**, an ingenious scientific-educational artificial satellite realized through a collaborative pact between the Russian Federal Space Agency and the Indian Space Research Organisation (ISRO). The satellite's foundation rested upon ISRO's Indian Mini Satellite-1 bus, encapsulating the amalgamation of technological prowess from both nations.

A pivotal moment materialized on the 20th of April 2011, when YouthSat embarked on its celestial journey through the propulsive force of the Polar Satellite Launch Vehicle. This launch site was the esteemed Sriharikota, India, underlining ISRO's commitment to pioneering endeavours and fostering international partnerships.

ISRO's voyage into space innovation extended with the launch of Jugnu, a remarkable Indian CubeSat satellite. As a technology demonstrator and remote sensing unit, Jugnu took flight on the 12th of October 2011. Operated by the Indian Institute of Technology (IIT) Kanpur, Jugnu underscored the synergy between educational institutions and cutting-edge space technology.

Jugnu Satellite

Concurrently, the horizon of innovation was enriched with SRMSAT, a Nanosatellite meticulously crafted by students at Sri Ramaswamy Memorial University (SRM Institute of Science and Technology, Chennai). Launched on the same day as Jugnu, SRMSAT embarked on a dual mission. Its primary objective centred on developing a robust nanosatellite platform for forthcoming ventures, while its secondary goal involved monitoring greenhouse gases. Characterized by a compact structure, SRMSAT boasted dimensions of 28 centimetres in length, height, and width, weighing 10.4 kilograms.

Jugnu and SRMSAT embarked on their cosmic journeys from the distinguished Satish Dhawan Space Centre at Sriharikota, further solidifying ISRO's legacy as a trailblazing force in space exploration. These endeavours underscored India's technological acumen and its commitment to fostering educational growth and forging international collaborations that continue redefining the boundaries of space exploration.

SRMSAT Satellite

SathyabamaSat: A micro experimental satellite emerged from the collaborative efforts of students and faculty at Sathyabama University, Chennai, Tamil Nadu. This ingenious creation gathered crucial data on greenhouse gases, including water vapour, carbon monoxide, carbon dioxide, methane, and hydrogen fluoride.

Swayam: A 1-U picosatellite tells a tale of innovation by undergraduate students from the College of Engineering, Pune, Maharashtra. This satellite pioneered the demonstration of passive attitude control—a technique for stabilizing and orienting the satellite—a maiden endeavour for an Indian satellite. Additionally, Swayam seeks to unravel the mysteries of low-earth-orbit channel characteristics in the UHF ham band.

Pratham: An Indian ionospheric research satellite, embodies the aspirations of the Indian Institute of Technology Bombay as part of the Student Satellite Initiative. This compact cube, with sides spanning 30 centimetres, carries a noble purpose—to count electrons in Earth's ionosphere. Beyond its scientific mission, Pratham is a beacon for knowledge and experience acquisition in Satellite and Space Technology. The project's essence lies in empowering the Satellite Team's nurturing skills across phases like Design, Analysis,

Fabrication, and Testing, culminating in creating the Flight Model. With every achievement, ISRO paves the path for an enlightened future in space exploration.

Space Telescope:

Astrosat, India's first dedicated multi-wavelength space telescope, was launched on a PSLV-XL on 28 September 2015. It became India's first dedicated Space Astronomy Observatory.

What sets Astrosat apart is its remarkable capability to conduct broad-spectrum multi-wavelength observations, stretching from the far reaches of ultraviolet light to the enigmatic realm of gamma rays. Among its stellar instruments, the Ultra Violet Imaging Telescope (UVIT) stands out with an unprecedented angular resolution of 1.5 arc seconds—three times superior to the operational GALEX-Galaxy Evolution Explorer. Meanwhile, the Large Area X-ray Proportional Counter (LAXPC) boasts the largest collecting area among X-ray detectors. At the same time, the Cadmium Zinc Telluride Imager (CZTI) possesses a unique capability to measure X-ray polarization and extends its reach beyond 100 keV.

ISRO's visionary approach extended to collaborating with prestigious Indian academic institutions. The Tata Institute of Fundamental Research (TIFR) Mumbai, Indian Institute of Astrophysics (IIA) Bangalore, Inter-University Centre for Astronomy & Astrophysics (IUCAA) Pune, and the Physical

Research Laboratory (PRL) Ahmedabad were critical enablers of Astrosat's mission.

The observatory's scientific journey pivots around the cornerstone of broadband spectroscopic investigations. Its instruments delve into the narratives of X-ray binaries, Active Galactic Nuclei (AGN), Supernova Remnants (SNRs), galaxy clusters, and the enigmatic coronae shrouding stars. Astrosat's mission extends to exploring the periodic and non-periodic variability inherent in X-ray sources, offering insights into the dynamic rhythms of these cosmic entities. It also Conducts low- to moderate-resolution spectroscopy across an expansive energy spectrum. Astrosat focuses on the profound study of X-ray-emitting objects. The observatory's instruments probe timing studies, unravelling the rhythmic dances within periodic and aperiodic phenomena within X-ray binaries and investigating the pulsations originating from X-ray pulsars.

Delving into the intricate variations showcased by X-ray binaries—quasi-periodic oscillations, flickering, flaring, and more—Astrosat enriches our understanding of their intricate dynamics. It delves into the realms of short- and long-term intensity fluctuations within active galactic nuclei, shedding light on the mechanisms orchestrating their luminosity variations. Through meticulous time-lag studies encompassing low/hard X-rays and UV/optical radiation, Astrosat unravels the temporal correlations threading through these radiant realms.

A remarkable facet of Astrosat's mission involves capturing the ephemeral brilliance of X-ray transients—potent cosmic phenomena. This endeavour highlights active galactic nuclei, suspected abodes of super-massive black holes. Astrosat attempts to unravel the mysteries surrounding these colossal cosmic entities by directing its instruments towards these enigmatic regions.

AstroSat's achievements radiate through its many accolades. UVIT's unparalleled angular resolution, LAXPC's expansive collecting area, and CZTI's remarkable capability to measure X-ray polarization testify to the observatory's innovation. AstroSat's data has uncovered UV photons from a redshift of 1.42, untangled the enigma of a cosmic source radiant in both the infrared and UV domains, and illuminated the X-ray polarization emerging from the off-pulse region of the Crab pulsar, among other remarkable findings.

AstroSat's legacy finds its voice through scholarly channels, resonating through over 275 articles published in refereed journals and more than 500 GCN circulars, Astronomer's Telegrams, and conference proceedings as of September 2022. Operating as an observatory-class telescope, AstroSat embodies the spirit of collaboration, inviting national and international users to join its cosmic odyssey. The call for proposals, typically released in the first quarter of each calendar year, beckons the global community to contribute to this cosmic symphony. At present, AstroSat proudly hosts close to 2,000 registered users from 54 countries.

Anti-Satellite Weapon - Mission Shakti:

India's space program plays a vital role in the country's security, economy, and social progress. It involves various spacecraft missions, such as communication, weather prediction, observing Earth, navigation, scientific research, and defence. Safeguarding these space assets is crucial, which led to the significant achievement known as Mission Shakti.

In 2006 and 2007, India successfully tested its initial exo atmospheric interceptor, Prithvi Air Defence (PAD) and the endo-atmospheric interceptor, Ashwin/Advanced Air Defence. In 2009, India developed a new exo atmospheric interceptor called Prithvi Defense Vehicle (PDV), akin to the Terminal High Altitude Area Defense (THAAD) system. The year 2014 marked India's first successful test of PDV. In subsequent years, real-time interception tests were conducted against manoeuvring targets, one in 2017 and another in 2019. In 2017, India encountered an issue with its critical imaging satellite, RISAT-1. Responding to this, the Indian Government initiated Project XSV-1 in 2016 for an Anti-Satellite (ASAT) test. They used a modified PDV called PDV MkII to test against a satellite.

On March 27, 2019, a momentous event occurred when India executed an anti-satellite (ASAT) missile test from the Dr. A.P.J. Abdul Kalam Island launch complex. Through this mission, India successfully showcased its ability to intercept a satellite in outer space using indigenous technology. The interceptor missile featured a Kinetic kill vehicle with hit-to-kill capabilities, making it a direct-ascent anti-satellite weapon. The test successfully struck a test satellite at an altitude of 283 km in low Earth orbit (LEO), marking Mission Shakti as a triumphant ASAT missile test.

The missile had a length of 13 meters and a diameter of 1.4 meters. It consisted of three stages with solid-propellant rocket motor stages and a Kill vehicle. With this capability, India joined an exclusive group of nations, including the USA, Russia, and China. The mission was led by the Indian Space Research Organisation (ISRO) and the Defence Research and Development Organisation (DRDO). The success of Mission Shakti demonstrated India's capacity to safeguard its space interests, adding a new dimension to warfare.

The ASAT test showcased the prowess of a modified anti-ballistic missile interceptor, Prithvi Defence Vehicle Mark-II, developed under Project XSV-1. India's accomplishment also hinted at its potential to intercept intercontinental ballistic missiles (ICBMs) and act as a deterrent. The Indian Ministry of External Affairs clarified that the test occurred at a low altitude to ensure that resulting debris would naturally decay and re-enter Earth's atmosphere within weeks.

India established the Defence Space Agency to enhance outer space protection capabilities and counter space-based threats. A simulated space warfare exercise named IndSpaceEx was conducted in July 2019 to assess potential threats and devise a joint space warfare doctrine. India is also working on advanced ASAT technologies like directed energy, co-orbital weapons, lasers, and electromagnetic pulse (EMP)-based weaponry. Collaborative efforts between ISRO and DRDO continue to drive the success of these missions.

Chandrayaan Program

The Chandrayaan project's announcement came from Prime Minister Atal Bihari Vajpayee during his Independence Day speech on August 15, 2003. This marked a significant leap for India's space endeavours. The concept of an Indian mission to the Moon was initially proposed in 1999 at a meeting of the Indian Academy of Sciences. The Astronautical Society of India furthered this idea in 2000. Subsequently, the Indian Space Research Organisation (ISRO) established the National Lunar Mission Task Force. This group determined that ISRO possessed the necessary technical prowess for an Indian lunar mission.

In April 2003, more than 100 distinguished Indian scientists specializing in planetary and space sciences, Earth sciences, physics, chemistry, astronomy, astrophysics, engineering, and communication convened. They approved the Task Force's recommendation to launch a Moon probe. By November of the same year, the Indian government approved the mission, setting the course for Chandrayaan's journey. The scientists and ISRO think-tank knew this would be a multiple-mission program. The program had three distinct phases. They are as follows.

- **Phase I:** Moon Orbiter & Impactor – this phase has been successful with mission Chandrayaan-1.
- **Phase II:** Soft landing on the moon and exploration of moon's surface via a rover. The Chandrayaan-2 did not accomplish the soft landing on the moon, ; hencehandrayaan-3 was launched to achieve this task.
- **Phase III:** On-site sampling - The next mission will be the Moon's Polar Exploration Mission, or ISRO would continue the naming convention and call it Chandrayaan-4, which is to be launched in a time frame of 2026-28.

Chandrayaan-1

Chandrayaan-1 was the first mission under the Chandrayaan Program. ISRO pushed the boundaries of space exploration with its most celebrated mission, Chandrayaan-1, which marked a significant step for India in space exploration and garnered global attention.

The Chandrayaan-1 mission was publicly announced by India's Prime Minister, Atal Bihari Vajpayee, during his Independence Day speech on 15 August 2003. This proclamation marked the beginning of a profound journey in India's space program, eventually leading to ground-breaking discoveries about our Moon.

On 22 October 2008, at 06:22 Hrs IST, Chandrayaan-1 was launched from the Satish Dhawan Space Centre using ISRO's PSLV C11 launch vehicle. The PSLV C11 is an impressive piece of engineering, standing 44.4 metres tall and consisting of four stages that alternately use solid and liquid propulsion systems. The first stage of this majestic rocket carries 138 tonnes of propellant and ranks among the world's largest solid propellant boosters. Augmenting the first stage are six solid propellant strap-on motors (PSOM-XL), each loaded with twelve tonnes of solid propellant. The subsequent stages are equally formidable: the second stage leverages 41.5 tonnes of liquid propellant, the third utilizes 7.6 tonnes of solid propellant, and the fourth is fitted with a twin-engine configuration that uses 2.5 tonnes of liquid propellant.

Chandrayaan-1's mission incorporated both a lunar orbiter and an impactor. Following the successful launch, the spacecraft was smoothly inserted into lunar orbit on 8 November 2008. The next significant milestone was reached on 14 November 2008, when the Moon Impact Probe separated from the Chandrayaan-1 orbiter at 14:36 UTC and deliberately struck the south pole. The impact point, near the crater Shackleton, was named Jawahar Point. This achievement positioned ISRO as the fifth national space agency to touch the lunar surface. Chandrayaan-1's mission was comprehensive and multifaceted. Over a planned two-year period, it aimed to produce a complete map of the lunar surface's chemical composition and three-dimensional topography. Particular interest was focused on the polar regions suspected of containing water ice.

The Chandrayaan-1 Spacecraft

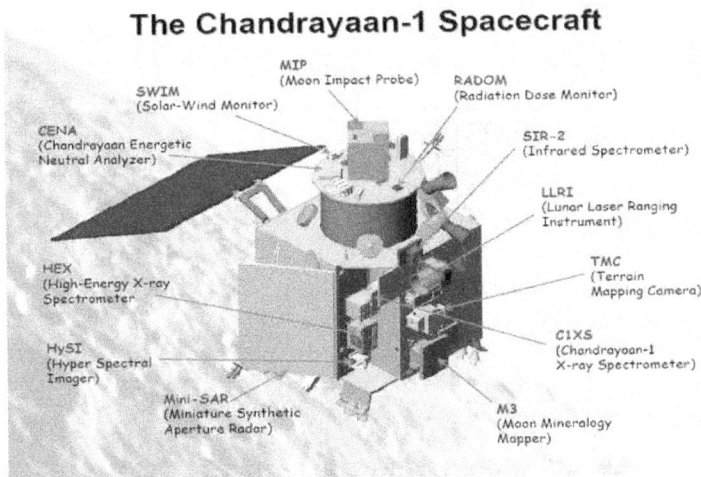

Below is the list of the various Indian and foreign payloads carried onboard Chandrayaan-1 to achieve the scientific objectives:

Scientific Objective	Payload	Agency
Chemical Mapping	HEX - High Energy X-ray Spectrometer	ISRO
	C1XS - Chandrayaan-1 X-ray Spectrometer	ESA
Mineralogical Mapping	HySI - Hyper Spectral Imaging Camera	ISRO
	SIR 2- Near Infrared Spectrometer	ESA
	M3 - Moon Mineralogy Mapper	NASA
Map and Characterize nature of Polar region	Mini - SAR Synthetic Aperture Radar	NASA
Topography Mapping	LLRI - Lunar Laser Ranging Instrument	ISRO
	TMC - Terrain Mapping stereo Camera	ISRO
Radiation Environment	RADOM - Radiation Dose Monitor Experiment	BAS
Magnetic Field Mapping	SARA - Sub Kev Atom Reflecting Experiment	ESA
Lunar Atmospheric Constituent	MIP - Moon Impact Probe	ISRO

ISRO - Indian Space Research Organisation

ESA - European Space Agency

NASA - National Aeronautics and Space Administration

BAS - Bulgarian Academy of Sciences

The mission led to remarkable achievements, notably the discovery of widespread water molecules in lunar soil. Collaborating with international agencies, NASA contributed two instruments: the Moon Mineralogy Mapper (M3) and the Miniature Synthetic Aperture Radar (Mini-SAR), both seeking ice at the poles. ISRO's principal instruments, including the Terrain Mapping Camera, HyperSpectral Imager, and Lunar Laser Ranging Instrument, produced highly detailed images and topographic maps of the lunar surface. In collaboration with the European Space Agency (ESA), the Chandrayaan-1 Imaging X-ray Spectrometer was designed to detect various minerals by the X-rays they emit when exposed to solar flares, employing the Solar X-Ray Monitor for part of this process.

The mission extensively mapped the Moon in infrared, visible, and X-ray light from lunar orbit, utilizing reflected radiation to explore various elements, minerals, and ice. One of the primary scientific objectives of Chandrayaan-1 was to confirm the presence of water on the Moon. This discovery was central to ISRO's plans and resonated globally, affecting future human settlements and theories about the Moon's origin. Chandrayaan-1's confirmation of lunar water significantly rejuvenated global interest in lunar exploration. Remarkably, the entire mission cost less than **$100 million**. The collaboration model, involving foreign instruments, greatly enhanced the scientific outcomes without inflating the budget.

Chandrayaan-1 significantly boosted India's space program, reflecting the country's ability to research, develop, and utilize indigenous technology to explore the Moon. It symbolized India's capabilities, technological advancements, and aspiration to contribute significantly to global space research.

Mangalyaan: Mars Orbiter Mission (MOM)

The Mars Orbiter Mission (MOM), colloquially called Mangalyaan, marked India's inaugural step into interplanetary exploration. MOM's primary objective revolved around an extensive investigation of Mars, including examining its surface features, morphology, mineralogy, and atmospheric composition. The dedicated search for methane within the Martian atmosphere was a distinctive facet of the mission. This quest could illuminate the possibility of past or existing life on the Red Planet.

Capitalizing on the expertise gleaned from Chandrayaan 1, India's first lunar orbiter, ISRO embarked on the development of Mangalyaan. The Martian spacecraft was a strategic evolution of the Chandrayaan 1 model, strategically augmented with upgraded components as necessitated by the mission's unique objectives.

Initially, ISRO envisaged launching Mangalyaan using their Geosynchronous Satellite Launch Vehicle (GSLV) rocket. The GSLV could propel the spacecraft beyond Earth's orbit onto an interplanetary trajectory towards Mars, a trajectory standard for most Mars missions. However, in a twist of fate, the GSLV faced two critical failures in 2010, coinciding with Mangalyaan's conceptualization.

The unforeseen challenges the GSLV's failures posed necessitated extensive design rectifications and prolonged preparations, estimating a timeline of around three years. This timeline would have collided with the time-sensitive launch window for Mars in November 2013, with the subsequent launch opportunity only available in 2016. Consequently, ISRO's pragmatic approach led them to opt for an alternative launch method. In 2013, Mangalyaan was successfully launched aboard the Polar Satellite Launch Vehicle (PSLV).

However, the PSLV's capabilities had limitations. While it could position Mangalyaan in a highly elliptical Earth orbit, the spacecraft had to take charge of its interplanetary trajectory. This intricate task involved firing the spacecraft's engines at precise intervals during each orbit over the following weeks. The objective was to set Mangalyaan on course for Mars, a trajectory divergent from conventional Mars missions. Yet, this innovative trajectory proved effective. Roughly 300 days after its launch, Mangalyaan executed another engine burn to manoeuvre itself into Mars' gravitational field, triumphantly achieving Mars orbit. The probe travelled a distance of 78,00,00,000 kilometres to reach Mars. ISRO was the world's first space agency to succeed in the Mars mission in the maiden attempt.

The Mangalyaan spacecraft achieved a momentous feat on September 23, 2014, as it seamlessly entered the orbit of Mars, elevating ISRO's status to that of the fourth global space agency to accomplish this extraordinary endeavour.

Initially intended for a 6-month mission life, the Mars Orbiter surpassed all expectations, culminating in a remarkable 7-year presence within Mars' orbit, ultimately marking September 24, 2021, as a historic milestone. India had even left China behind in this feat because China and Japan failed their Mars mission.

Mars Orbiter

The mission's financial outlay was an impressive ₹450 Crore (equivalent to US$73 million), positioning it as the most cost-effective Mars exploration mission to date. This economic efficiency could be attributed to a meticulously devised "modular approach," streamlined ground testing procedures, and the dedicated commitment of scientists who tirelessly dedicated 18 to 20 hours daily to the mission's success.

ISRO's ground-breaking achievement of placing a spacecraft in Mars' orbit on its maiden attempt garnered global admiration and recognition. Notably, within India, the mission's impact was even more profound, bolstered by ISRO's pioneering use of social media to disseminate awareness about the endeavour on a larger scale.

The mission's triumph inspired numerous adaptations within India's cultural landscape, including captivating film and television depictions, with "Mission Mangal" standing out as the most prominent. The Indian government also made a distinctive decision, featuring an illustration of the Mangalyaan spacecraft on the reverse side of the nation's highest denomination currency note, the ₹2,000 bill. Additionally, Minnie Vaid penned a noteworthy literary work titled "Those Magnificent Women and their Flying Machines," which artfully profiles the remarkable journeys of pivotal women who played instrumental roles in the mission's realization.

Real MOMs in ISRO.

A photograph captured the jubilant faces of women, initially believed to be scientists, shattering the stereotype that rocket science in India was an exclusively male domain. While ISRO clarified that these were administrative staff, it proudly affirmed the presence of numerous women scientists who significantly contributed to missions and graced the control room during launches.

In an industry where gender disparity persists, ISRO is a testament to women's remarkable achievements. VR Lalithambika, a control systems engineer, was pivotal in the agency. Muthayya Vanitha and Ritu Karidhal led the Chandrayaan-2 mission as project and mission directors, making Vanitha the first female project director in ISRO. Karidhal, nicknamed "Rocket Woman," was a driving force behind the Mangalyaan and Chandrayaan-3 missions. N Valarmathi, a seasoned scientist with three decades at ISRO, directed the indigenously developed Radar Imaging Satellite, RISAT-1, following in the footsteps of Anuradha, the second woman scientist to lead such a prestigious project. Seetha Somasundaram, Program Director at ISRO, orchestrated space science endeavours, while Nandini Harinath managed Mars mission designs. As a Scientist and engineer, Minal Rohit spearheaded the Methane Sensor for MOM, and Anuradha TK served as the Geosat Programme Director at ISRO Satellite Centre.

Gender equality thrives at ISRO, where recruitment and advancement hinge on merit and contributions. Here, being a woman neither grants special privileges nor subjects one to discrimination; equality reigns. Amidst ISRO's soaring triumphs, these women illuminate the realm of space science, solidifying their place as Real MOMs—pioneers who break barriers, shatter ceilings, and inspire generations.

Cartoon Controversy

As India and ISRO basked in international congratulations for their milestone achievement, an unexpected incident unfolded when The New York Times published an article titled "India's Budget Mission to Mars," accompanied by a controversial cartoon by Heng Kim Song. The cartoon depicted a destitute farmer with a cow knocking at the door of the "Elite Space Club," where two men casually read a newspaper heralding India's space feat.

The cartoon's intent could not be concealed, and the veil of elitism was transparent. The imagery showcased a barefooted man dressed in traditional attire, donning a dhoti and turban, accompanied by a cow – a portrayal seemingly reminiscent of India's colonial past, marked by British rule and the height of racial prejudices. The cartoon's undertones were unmistakable, raising concerns about its elitist undertones and, more disturbingly, its racial connotations.

The condemnation of The New York Times' actions reverberated across the globe. Criticism flowed from international news outlets, including those in the United States and netizens worldwide, who quickly identified the offensive elements within the article and cartoon. Consequently, the editorial page editor of The New York Times extended an apology, acknowledging the insensitivity and distress caused by their misstep.

This incident serves as a reminder of journalism's broader implications and responsibilities, emphasizing the need for sensitivity, cultural awareness, and a commitment to ethical reporting, especially when dealing with matters that transcend national and racial boundaries.

ISRO's Global Recognition

In the realm of global space exploration, there are a total of 77 space agencies in various countries all around the world, out of which 16 space agencies even have the capacity to launch a satellite. Notably, the European Space Agency (ESA) commands a multinational coalition of 22 European nations, with additional agencies rapidly emerging. This landscape also witnesses the rise of private space agencies, exemplified by entities like SpaceX and Blue Origin, often founded with a focus on ventures such as space tourism.

Among all these space agencies, the Indian Space Research Organisation (ISRO) has consistently demonstrated its prowess as a cost-effective and reliable entity for satellite launches. This track record has garnered global recognition and trust, positioning India as an esteemed choice on the world stage. As the commercialization of space transportation gains traction, India stands advantageously poised for the future.

ISRO's achievements span an impressive portfolio, successfully launching 431 satellites on behalf of 34 countries. This roster includes nations like Germany, South Korea, Belgium, Indonesia, Argentina, Italy, Israel, the Netherlands, Denmark, Turkey, Switzerland, Algeria, Norway, Luxembourg, Singapore, Austria, the United Kingdom, Kazakhstan, the United Arab Emirates, Chile, the Czech Republic, Finland, Latvia, Lithuania, Slovakia, Colombia, Malaysia, Spain, Brazil, Mexico, Canada, Australia, and more.

ISRO's capacity to attract even countries with well-established space agencies sets it apart. Notable examples include China, the United States, Japan, and France, which opt for ISRO to launch their satellites despite possessing their own launch capabilities. This trend underscores the immense trust in ISRO's launch proficiency and demonstrates its cost-effectiveness. This choice suggests confidence in ISRO's abilities that even their space agencies or launch vehicles cannot surpass.

ISRO's journey is emblematic of India's ascent in the global space arena, characterized by innovation, reliability, and affordability. As the world continues to embrace space exploration and utilization, ISRO's legacy remains steadfast, shaping the narrative of space initiatives with each successful launch and forging a path that beckons the stars for future generations.

Foreign satellites launched by ISRO:

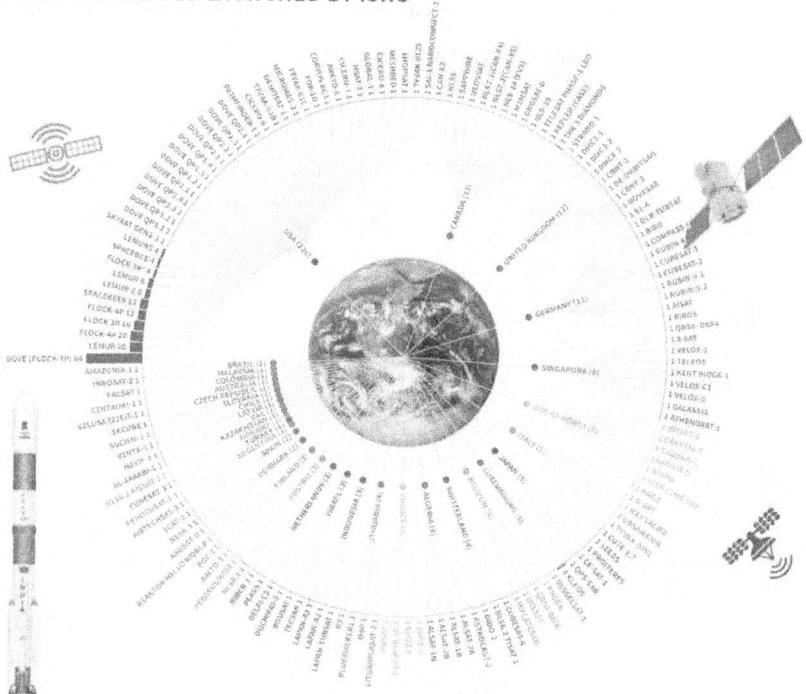

FOREIGN SATELLITES LAUNCHED BY ISRO

India's prowess in achieving significant space missions on modest budgets is a testament to its innovative strategies and simplicity. Guided by a visionary perspective, Indian leadership has consistently envisioned space as a catalyst for national development. Given India's economic challenges as a newly independent nation, the space program was strategically designed to be innovative and cost-efficient, with a resolute focus on practical applications.

A compelling example of this approach is the Mars mission. ISRO succeeded with a budget of Rs 800 crore, remarkably undercutting the Hollywood blockbuster "Martian," which incurred Rs 1,062 crore in expenses. Similarly, the triumph of the unmanned spacecraft Mangalyaan reaching Mars in 2014 for $74 million stood in stark contrast to NASA's expenditure of $671 million on comparable missions. Notably, this achievement surpassed the budget of the acclaimed film "Gravity." India's investment was less than one-tenth of the US Mars mission Maven's cost.

ISRO's commitment to indigenization remains integral to its strategies. Collaborating with the industry, ISRO works diligently to develop critical components and materials domestically, reducing reliance on imports. The Indian sector plays a pivotal role in the entire process, contributing extensively to designing, manufacturing, and testing components and sub-systems in alignment with ISRO's specifications.

104 satellites on one Rocket:

After achieving significant milestones with Chandrayan-1 and Mangalyaan missions, ISRO broke new ground in space exploration. On 15 February 2017, ISRO's heavy-duty launcher - PSLV-XL, took to the skies from the Satish Dhawan Space Centre in Sriharikota. Its mission is to successfully deploy a staggering 104 satellites into sun-synchronous orbits in one go. This surpassed the previous record set by Russia's Dnepr rocket, which had launched 37 satellites on 19 June 2014.

The primary payload of this historic launch was the Cartosat-2D earth observation satellite. Accompanying it were 103 nanosatellites, two of which were ISRO experiments. One hundred nanosatellites were part of a commercial deal between global firms and ISRO's commercial arm, Antrix Corporation Limited.

Deploying such a vast number of satellites necessitated a meticulously planned release strategy spanning three stages. Initially, the three larger Indian satellites unfurled axially aligned with the vehicle. This was succeeded by the radial release of 81 nanosats, moving away from the rocket. The mission culminated with the sequential separation of the final 20 nanosats. A mere 16 minutes and 48 seconds post-launch, Cartosat-2D was positioned at approximately 510 kilometres altitude, swiftly followed by ISRO's nanosatellites, INS-1A and INS-1B. PSLV C-37 then spent another 11 minutes ensuring the remaining 101 co-passenger satellites reached their designated orbits.

The enormity of this achievement did not go unnoticed. India's Prime Minister, Narendra Modi, lauded ISRO, tweeting, "This remarkable feat by @isro is yet another proud moment for our space scientific community and the nation. India salutes our scientists." Even former US Senator Dan Coats was taken aback, commenting, "I was shocked the other day to read that India, on one rocket launch, deposited more than 100 satellites in space."

Cartoon Controversy Reply:

Following ISRO's Mangalyaan mission triumph, The New York Times took a jab at the accomplishment through a provocative cartoon. Although ISRO chose not to respond officially, anticipation brewed among Indians and global netizens for a fitting retort. In an astonishing turn of events merely three years later, ISRO presented a perfect chance. ISRO launched 104 satellites into orbit in a groundbreaking feat, setting a world record. Seizing the moment, netizens reciprocated the Times' cartoon in jest.

Earlier, the caricature from The New York Times had portrayed a destitute farmer with a cow knocking at the door of the "Elite Space Club." Inside, two men casually perused a newspaper, celebrating India's celestial achievement. The retort cartoon featured the same farmer and his cow now seated within the elite space club. They read a newspaper heralding ISRO's record-breaking launch of 104 satellites while the elites stood outside, clutching their rockets, eager to gain entry.

Twitter and other social platforms reverberated in the following weeks with diverse reactions. While some conveyed congratulations and pride, others took playful jabs at neighbouring nations, particularly Pakistan. Humorous memes and animated gifs inundated social media channels, with a few even drawing parallels between ISRO's achievements and those of NASA.

Frank Sinatra's words, "The best revenge is a massive success," resonated deeply. ISRO exemplified this adage, showcasing remarkable resilience and triumph. Through their unparalleled achievements, they silenced sceptics and etched their legacy on the global space exploration stage.

Geosynchronous Satellite Launch Vehicle (GSLV)

The Geosynchronous Satellite Launch Vehicle (GSLV) project began in 1990, enabling India to launch geosynchronous satellites independently. This launch vehicle was designed by the Indian Space Research Organization (ISRO) and utilizes proven components from the Polar Satellite Launch Vehicle (PSLV). The GSLV's third stage is powered by a LOX/LH2 Cryogenic engine, crucial for propelling satellites into geostationary transfer orbits (GTO).

On April 18, 2001, the GSLV's first developmental flight (Mk I configuration) encountered failure, unable to place the payload in the intended orbit. However, after a successful launch of the GSAT-2 satellite in its second developmental flight, the GSLV was declared operational. The launcher faced challenges initially, achieving only two successful launches out of 7 from its debut until 2014.

Standing 49 meters tall and weighing 415 tons, the GSLV is a three-stage vehicle incorporating solid, liquid, and cryogenic stages. With a lift-off mass of 408 long tons, it can deploy payloads of up to 5,000 kg in a low Earth orbit (LEO) or 2,500 kg (for the Mk II version) in an 18° geostationary transfer orbit.

The GSLV comprises three stages with strap-on motors. The first stage (GS1) employs the S139 solid rocket motor, generating a thrust of 4700 kilo Newton, while the second stage (GS2) features the Vikas liquid rocket engine producing 800 kilo Newton of thrust. The third stage (CUS) employs a Cryogenic engine powered by liquefied oxygen and hydrogen, a technological feat to end India's reliance on foreign cryogenic engines. The CE-7.5 is India's first indigenous cryogenic engine, capable of generating a thrust of 75 kilo Newton.

GSLV-D5's launch on January 5, 2014, marked a significant milestone, as it successfully utilized the indigenous CE-7.5 cryogenic engine. The GSLV's current configuration with CE-7.5 allows it to carry payloads of up to 2,500 kilograms to the Geostationary Transfer Orbit (GTO) or payloads of up to 5 tonnes in Low Earth Orbits (LEO).

Interestingly, GSLV has been rebranded as LVM-3, denoting Launch Vehicle Mark 3. This name change reflects its new purpose – not deploying satellites into geosynchronous orbits at an altitude of 35,786 kilometres above Earth's equator. This strategic shift showcases ISRO's adaptability and ongoing commitment to advancing space exploration and technology.

GSLV Variants:

IS RO has developed several variants of the Geosynchronous Satellite Launch Vehicle (GSLV), including GSLV Mk-I, GSLV Mk-II, and GSLV Mk-III. The first variant, GSLV Mk I, is no longer active in service. GSLV underwent different configurations during its evolution, designated as Mk I a, b, and c, and Mk II. The cryogenic third stage of GSLV Mk.1 utilized the Russian KVD-1M engine. This engine, initially designed for the cryogenic version of the Proton launch vehicle, was later replaced by the Indian-built CE7.5 engine for the Mk.2 version.

In the retired GSLV Mk-Ia configuration, the GSLV employed a 125-ton Core Stage along with a Russian-made Cryogenic Upper Stage. The Indian cryogenic stage was not ready at that time. The first developmental flight of Mk-Ia took place on April 18, 2001.

While it marked the maiden launch of the GSLV class, the mission only achieved partial success. The payload, GSAT-1 Communications Satellite, ended up in a lower-than-planned orbit due to a performance shortfall caused by the vehicle's guidance system or a premature shutdown of the third stage. GSAT-1's propulsion system flaw hindered it from reaching Geostationary Orbit, rendering it unusable for its intended purpose.

Geosynchronous Satellite Launch Vehicle Mark II (GSLV Mk II) is another ISRO creation designed to launch communication satellites into geostationary transfer orbit using a cryogenic third stage. Initially, Russian GK-supplied cryogenic stages were utilized, but they were later replaced with the domestically developed cryogenic stage, incorporated from GSLV D5 onwards, beginning in January 2014. This fourth-generation launch vehicle boasts three stages and four liquid strap-ons. The Cryogenic Upper Stage (CUS), indigenously developed and flight-proven, forms the third stage. GSLV Mk II has achieved six consecutive successful flights since 2014.

The vehicle's first stage is the S139 solid-fueled stage, also used in the Polar Satellite Launch Vehicle (PSLV). Four liquid-fueled strap-on boosters with Vikas engines surround the core. The second stage features a single modified Vikas engine, utilizing storable propellants. The third stage operates on liquid Oxygen and liquid Hydrogen, consumed by an ICE engine, marking GSLV's cryogenic stage.

GSLV can deploy payloads of up to 2,500 kilograms to Geosynchronous Transfer Orbit, while its Low Earth Orbit capability is 5,000 kilograms. The vehicle's operations are centred at the Satish Dhawan Space Center. The GSLV series stands as a testament to ISRO's innovative approach and progress in mastering complex launch technologies.

Launch Vehicle Mark-3

The GSLV Mark-III, now rebranded as Launch Vehicle Mark-3 (LMV3), is a distinct and robust launch vehicle. This new heavy-lift launch vehicle ushers in a cost-effective approach, enabling the launch of 4000 kg spacecraft to Geosynchronous Transfer Orbit (GTO).

Following a sub-orbital test flight on December 18, 2014, ISRO triumphantly marked the first orbital test launch of LVM3 on June 5, 2017, from the Satish Dhawan Space Centre. This accomplishment set the stage for significant missions, including CARE, India's space capsule recovery experiment module, and the ambitious Chandrayaan-2 and Chandrayaan-3 missions. LVM3 is poised to further elevate its significance by carrying Gaganyaan, India's maiden crewed mission, as part of the Indian Human Spaceflight Programme.

LVM3 is meticulously configured as a three-stage vehicle. It features two robust solid strap-on motors, S200, and a decisive liquid core stage, L110, accompanied by a high-thrust cryogenic upper stage, C25. The S200 solid motor, one of the largest in the world, houses a whopping 204 tonnes of solid propellant. The liquid L110 stage, with twin liquid engines and 115 tonnes of liquid propellant, provides exceptional thrust. The C25 Cryogenic upper stage boasts an indigenous high-thrust cryogenic engine (CE20) with 28 tons of propellant loading. The vehicle's stature is 43.5 meters in length, with a gross lift-off weight of 640 tonnes, topped by a 5m-diameter payload fairing.

The inaugural flight of LVM3 on June 5, 2017, heralded a new era. This heavy-lift vehicle is a game-changer for ISRO, with a significant lift-off mass of 640 tons, enabling the launch of communication satellites weighing up to 4000 kg into GTO. The lift-off involves simultaneous ignition of the two S200 boosters, followed by the core stage (L110) ignition around 113 seconds into the flight. The S200 stages burn for approximately 134 seconds, separating them at 137 seconds. Payload fairing separation occurs at an altitude of 115 km, roughly 217 seconds into L110 firing. L110 burnout, separation, and C25 ignition synchronize at 313 seconds. The spacecraft is then injected into a Geosynchronous Transfer Orbit (GTO) orbit of 180x36000 km, reaching this milestone around 974 seconds.

LVM3's operational journey commenced on July 22, 2019, when it carried out the Chandrayaan-2 mission. This mission, involving an orbiter, lander, and rover, marked India's second venture to the Moon. Notably, the Chandrayaan-2 stack represented the heaviest spacecraft ever launched by ISRO.

On October 22, 2022, a significant milestone was achieved as LVM3 embarked on its first commercial launch. This marked India's entry into the global market for heavier payloads. It was also a debut for LVM3's polar low earth orbit launch and its first multi-satellite mission, carrying an impressive payload of about 6 tons.

As LVM3 continues to redefine India's capabilities in space exploration, it successfully propelled the Chandrayaan-3 mission, solidifying its role in ISRO's journey of innovation and achievement.

CARE

CARE, an abbreviation for Crew Module Atmospheric Re-entry Experiment, is more than just words—it embodies ISRO's quest to conquer new frontiers. This mission serves as a testing ground for re-entry technologies vital for the Crew Module, including validating a parachute-based deceleration system. The underlying purpose of CARE goes beyond its acronym, seeking to deepen our understanding of blunt body re-entry aerothermodynamics and the intricacies of parachute deployment in a cluster configuration.

The crew module, a cornerstone of this venture, was carefully nestled inside the payload fairing of the LVM3, showcasing the meticulous planning of ISRO's engineers. Crafted from aluminium alloy, CARE bore a lift-off mass of 3,735 kg, with a diameter of 3100 mm and a height of 2698 mm. The module's design featured an ablative thermal protection system, with Medium Density

Ablative (MDA) tiles adorning its side panels and a forward heat shield crafted from carbon phenolic tiles.

In anticipation of the grand experiment, a practice run for the recovery of the crew module was conducted on October 31, 2014, involving the Indian Coast Guard ship ICGS Samudra Pahredar. The stage was set for CARE's true voyage—a date etched in history as December 18, 2014. The experimental mission took flight aboard an LVM3, designated by ISRO as the LVM 3X CARE mission. As the crew module soared, it was gracefully separated at a planned altitude of 126 km and a velocity of 5300 m/s, heralding the crucial re-entry phase.

With careful orchestration, the Crew Module embarked on a journey that would put its systems to the ultimate test. Re-entering Earth's atmosphere at around 80 km, it then descended further, transitioning into a ballistic trajectory. Beyond 80 km, as the module followed an uncontrolled path, a decisive impact awaited at sea—approximately 180 km from the Andaman and Nicobar Islands. Here, the vigilant Indian Coast Guard stood ready for the module's retrieval.

After the re-entry saga, the Crew Module unveiled a grand spectacle—descent and splashdown. This sequence saw a comprehensive parachute system validation, showcasing apex cover separation and parachute deployment in a cluster configuration. The intricate deployment dance began as CARE slowed to 233 m/s. Successively, pilot parachutes, drogue parachutes, and main parachutes came into play, each pair contributing to a gradual deceleration. These main parachutes, boasting a 31-meter diameter, were considered India's most enormous parachutes ever used in any mission.

The final act unfolded as CARE gracefully touched down in the Bay of Bengal, around 600 km from Port Blair in the Andaman Islands and a considerable 1600 km from the Sriharikota launch site. As a testament to precision, the main parachutes detached, signalling the end of the experiment's dynamic phase. The Indian Coast Guard swiftly sprung into action, recovering CARE after tracking its beacon. The entire odyssey painted an awe-inspiring canvas spanning 20 minutes and 43 seconds, from launch to splashdown.

The significance of this mission reverberates beyond the ocean's surface. CARE's accomplishments pave the way for ISRO's audacious next step—the

Gaganyaan mission. ISRO charts its course toward an exhilarating future of space exploration through these resounding achievements.

Chandrayaan-2

The success of Chandrayaan 1 and the Mars Orbiter Mission (MOM) has boosted the confidence of ISRO's scientists. They planned phase II of the Chandrayaan Program, a more advanced soft landing mission on the Moon with Chandrayaan-2. This complex mission comprised an orbiter, lander, and rover. Its goal was to softly land on the Moon's unexplored South Pole and investigate its topography, minerals, atmosphere, and more.

Chandrayaan 2 was launched on July 22, 2019, from the same launch pad Chandrayaan 1 had taken off in Sriharikota. Instead of using the tested and reliable PSLV rocket used earlier, the spacecraft used the advanced Geosynchronous Satellite Launch Vehicle Mark III (GSLV Mk III). The mission marked a technological leap for ISRO. Its planned orbit around Earth had a perigee (nearest point to Earth) of 169.7 km and an apogee (farthest from Earth) of 45,475 km.

After launch, a series of manoeuvres raised its orbit, and by August 14, 2019, it escaped Earth's orbit to reach the Moon's vicinity. It was then inserted into lunar orbit on August 20, 2019. While orbiting the Moon at 100 km altitude, the Vikram Lander separated from the orbiter on September 2, 2019, preparing for landing. De-orbit manoeuvres adjusted Vikram's path, and began circling the Moon at 100 km x 35 km altitude. The descent of Vikram Lander went according to plan until it reached a height of 2.1 km.

The components of the Chandrayaan-2 were as below.

Chandrayaan-2 Orbiter:

The Chandrayaan-2 Orbiter Craft is designed with a cuboidal structure that accommodates various components. At one end are propulsion tanks and the launch vehicle's separation mechanism, while the other houses the lander. The Orbiter's decks contain systems that manage different functions of the spacecraft. Two solar panels, stored during launch, unfold after separation to supply power during different phases around Earth and the Moon. A Lithium-ion battery supports power during eclipses and peak requirements.

The Orbiter was a stable spacecraft with three-axis body stabilization and reaction wheels. Thrusters manage momentum and adjust its orientation. A liquid engine is used to move from Earth's parking orbit to a 100 km lunar orbit. The attitude and orbit control electronics gather data from star sensors and gyroscopes for spacecraft control. Sun sensors and accelerometers are other sensors employed.

The telemetry system relays health information, while the telecommand system handles command execution and distribution. The Orbiter's payloads are connected to the baseband data handling system for formatting and recording in a solid-state recorder for later playback. The RF system comprises an S-band TTC transponder and an X-band transmitter for transmitting payload data to the Indian Deep Space Network (IDSN) station. This data is sent through an X-band dual gimbal antenna directed towards the ground station.

The instruments on the Orbiter and their science objectives are as below:

Science Objective	Payload	Source
Chemical Mapping	CLASS - Large Area Soft X-ray Spectrometer	ISRO Sattelite Center, Bengaluru
	XSM - Solar X-ray Monitor	Physical Research Laboratory, Ahmedabad
Mineralogical Mapping	IIRS - Imaging IR Spectrometer	Space Applications Center (SAC), Ahmedabad
Map Lunar Exosphere	chACE 2 - Neutral Mass Spectrometer	Space Physics Laboratory, Thiruvananthapuram
Topography Mapping	TMC 2 - Terrain Mapping stereo Camera	Space Applications Center (SAC), Ahmedabad
Probe Lunar Surface for Water Ice, constituents	SAR - L & S band Synthetic Aperture Radar	Space Applications Center (SAC), Ahmedabad

The Chandrayaan-2 Orbiter is still in moon orbit and is used for Chandrayaan-3 mission.

Vikram Lander:

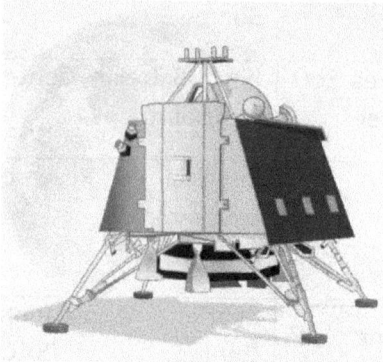

The lander for Chandrayaan-2 was named Vikram, honouring Vikram Sarabhai, the visionary behind India's space endeavours. Vikram's structure takes the shape of a truncated pyramid encasing a cylinder. Inside is room for the propellant tank and the interface connecting to the Orbiter's separation mechanism. Vertical panels hold solar cells, while stiffener panels house electronic systems. Four legs on the lander ensure stability upon landing on diverse terrains.

For power, body-mounted solar panels supply energy to systems throughout the mission phases—a lithium-ion battery step in during eclipses and the Lander's descent. Control electronics manage sensor and actuator interfaces. Sensors enable inertial navigation from separation to rough braking's end, and absolute sensors gauge position and velocity relative to the landing site. This guides the Lander from uneven braking to the designated landing spot. From separation onwards, the Lander takes charge of navigation, guidance, and control, ensuring a precise and safe soft landing on the Moon.

Four liquid engines generate braking thrust, while eight thrusters maintain the Lander's attitude. The leg mechanism absorbs energy upon touchdown, securing stable conditions for payload deployment and lunar science. Each leg features a telescopic assembly with crushable damping material and a foot pad. Rigorous tests ensure leg stability under harsh terrain conditions and high velocities.

Communication between the Lander and IDSN occurs in the S-band for telemetry and telecommand. High torque dual gimbal antennas transmit payload data. The Lander boasts a TM-TC data handling system with built-in storage. The Chandrayaan-2 Rover stays within the Lander during launch. Once landed, ramps deploy, and the Rover begins its lunar journey. Lander payloads will be deployed during the landing process.

The Vikram Lander is equipped with the following scientific equipments.

Science Objective	Payload	Source
Measure near surface plasma density and its changes with time	RAMBHA - Radio Anatomy of Moon Bound Hypersensitive Ionosphere and Atmosphere	Space Physics Laboratory, Thiruvananthapuram
Measure thermal properties of lunar Regolith near polar region	ChaSTE - Chandra's Surface Thermo Physical Experiment	Physical Research Laboratory, Ahmedabad Space Physics Laboratory, Thiruvananthapuram
Measure Lunar Seismicity around the landing site	ILSA - Instrument for Lunar Seismic Activity	Indian Space Research Organisaiton (ISRO)

Pragyan Rover:

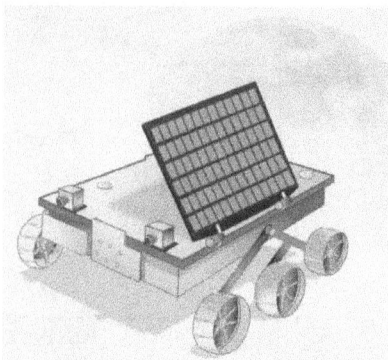

The Rover in the mission was named Pragyan, which translates to "wisdom" in Sanskrit. Weighing 27 kg (60 lb), it operated on solar power. Pragyan was designed as a six-wheeled mobility system, aiming to navigate the Moon's low gravity and vacuum environment. Alongside mobility, its purpose included scientific exploration of lunar resources. The Rover's design was inspired by NASA's "Sojourner," a space rover that explored Mars in 1997. Essential electronics were housed in the rover chassis, complemented by two navigation cameras generating stereo images for path planning. The solar panel provided the necessary power, while the rocker bogie mechanism and six wheels ensured robust mobility over obstacles and slopes on the exploration path.

Communication was facilitated via the Lander as Pragyan interacted with the Indian Deep Space Network (IDSN). The rover's two payloads conducted scientific investigations on the lunar surface. For navigation, Pragyan employed stereoscopic camera-based 3D vision. Two 1-megapixel monochromatic navcams in front of the rover offered the ground control team a 3D terrain perspective—this aided path planning by creating a digital elevation model of the surroundings. IIT Kanpur developed subsystems for light-based map generation and motion planning. Control and motor dynamics were managed by a rocker-bogie suspension system and six wheels powered by independent brushless DC electric motors. Steering was accomplished by varying the wheel speeds, a technique known as differential steering or skid steering.

Pragyan was anticipated to operate for one lunar day, equivalent to about 14 Earth days. Its electronics were not designed to endure the freezing lunar night. However, a solar-powered sleep/wake-up cycle could have extended its service time beyond the initial plan.

Chandrayaan-2 Rover has the below equipments on board.

Science Objective	Payload	Source
Study the elemental composition of lonar rock & soil	APXS - Alpha Particle X-ray Spectrograph	Physical Research Laboratory, Ahmedabad
Elemental analysis of lunar Regolith	LIBS - Laber Induced Breakdown Spectrograph	Laboratory for Electro-Optic Systems, Bengaluru

Loss of Vikram:

On the night of 6 September 2019, the Vikram lander began its descent towards the Moon at precisely 01:08:03, according to Indian Standard Time. The landing was scheduled to occur at around 01:23 a.m. However, the descent went differently than planned. As the spacecraft approached the lunar surface, ISRO's Telemetry Tracking and Command Centre lost communication with Vikram when it was merely 335 metres (0.335 km) above the Moon's surface.

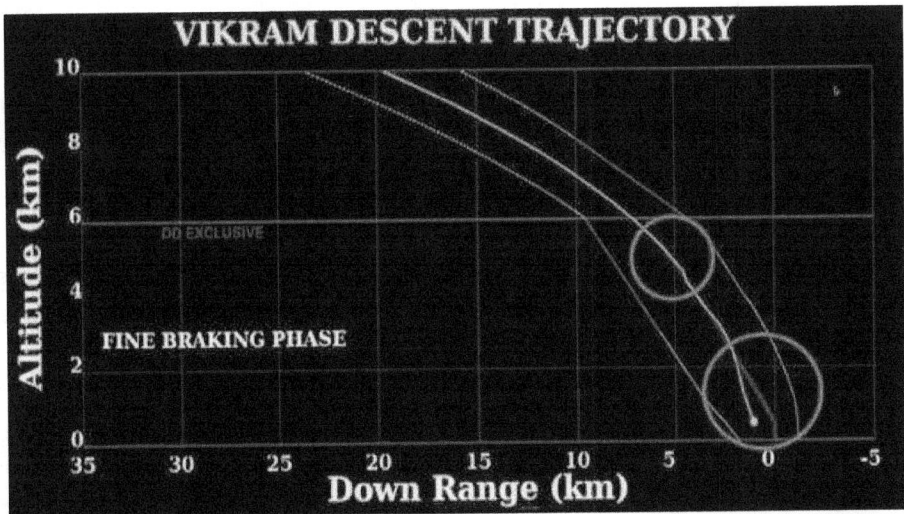

According to the available data, the issues began during the "Fine braking phase," initiated when Vikram was about 5 km above the Moon. This phase, taking the lander from an altitude of 5 km to 400 m, is crucial for a controlled descent. Unfortunately, anomalies started when Vikram was just over 2 km in altitude. The lander began deviating from its intended trajectory, and this deviation persisted as it dropped below 1 km altitude, even reaching near or below 500m. With the lander so close to the lunar surface, its speed was alarmingly high. It registered a vertical velocity of 59 metres per second (212 km/hr) and a horizontal velocity of 48.1 m/sec (173 km/hr). As a result, instead of nearing its predestined landing point, Vikram Lander was about 1.09 km away.

The original plan was for Vikram to decelerate considerably by the time it was 400m from the Moon's surface. It should have been almost hovering, ready to touch down at a "walking pace softly." Instead, a combination of unexpected challenges caused it to crash into the Moon.

The five engines designed to reduce the lander's velocity generated more thrust than anticipated. This led to the second challenge, where the accumulated errors exceeded the team's expectations. The software struggled to cope and was not designed for such high turn rates. Furthermore, a third issue arose: with its designated landing spot still distant, the lander's velocity unexpectedly increased. This was attributed to the chosen landing area being a rather tight 500 m x 500 m patch.

Ideally, Vikram was engineered to touch down at a speed of around 7 km per hour. However, owing to these complications, it's believed that Vikram made its unintended final impact at a velocity close to 50 m/s (180 km/h), a drastic difference from the intended two m/s (7.2 km/h). The Vikram lander software's capability to manage parameter dispersions was severely restricted.

Post-impact, both ISRO and NASA made intense attempts to establish communication with the lander. These efforts persisted for approximately two weeks before the harsh lunar night started. A couple of months later, on 16 November 2019, the Failure Analysis Committee presented its findings to the Space Commission. The primary reason for the unfortunate crash was a software glitch. The Lunar Reconnaissance Orbiter Camera (LROC) team pinpointed Vikram's impact site at coordinates 70.8810°S 22.7840°E. The aftermath of the crash was evident, with debris from the spacecraft scattered over multiple locations, covering several kilometres.

Aftermath:

There was an outpouring of support for ISRO from various quarters in the aftermath of the crash landing of its lunar lander. Various space agencies all over the globe congratulated India and ISRO on the Chandrayaan 2 attempt. American SpaceX CEO Elon Musk congratulated ISRO and encouraged ISRO to continue its work.

Below are some tweets for ISRO:

"In life, there are ups and downs. The country is proud of you. And all your hard work has taught us something ... Hope for the best ... You have served the country, science, and humanity well," said Prime Minister Narendra Modi. "As important as the final result is ... I can proudly say that the effort was worth it, and so was the journey. Our team worked hard, travelled far, and those teachings will always remain with us," Modi said in a speech posted on Twitter.

Space is hard. We commend @ISRO's attempt to land their #Chandrayaan2 mission on the Moon's South Pole. You have inspired us with your journey, and we look forward to future opportunities to explore our solar system together. – NASA, USA.

We are proud of India and its scientists today. Chandrayaan-2 saw some challenges at the last minute, but the courage and hard work you have shown are historical. Knowing Prime Minister @narendramodi, I have no doubt he and his ISRO team will make it happen one day. — Prime Minister of Bhutan.

Although it was not a successful landing this time, the world would recognise the significant technological advancement of the Indian Spacial Programme. We look forward to collaborative efforts between Mauritius and the ISRO team in the future. — Prime Minister of Mauritius.

We congratulate @ISRO on their incredible efforts on #Chandrayaan2. The mission is a huge step forward for India and will continue to produce valuable data to fuel scientific advancements. We have no doubt that India will achieve its space aspirations. AGW - Bureau of South and Central Asian Affairs (SCA)

Ambassador of Israel to India Ron Malka boosted the ISRO's confidence by saying, "Take pride India and have courage ISRO. This is a great achievement, and it is not the end. Israel is no stranger to the stumbling blocks on the way to a soft landing, and we know India will try again and complete the last step. We will see you there."

"Today marks a chapter in the great story of your space program. We look forward to the next one, and for a successful Chandrayaan-2 orbiter mission," the Planetary Society tweeted of the USA.

British newspaper The Guardian, in its article titled -- India's moon landing suffers last-minute communications loss, quoted Mathieu Weiss, a representative in India for France's space agency CNES, as saying, "India is going where probably the future settlements of humans will be in 20 years, in 50 years, 100 years."

An astronaut from Pakistan, Namira Salim, congratulated India on its historic attempt to make a lunar landing. Namira Salim is Pakistan's first astronaut. "The Chandrayaan-2 lunar mission is indeed a giant leap for South Asia which not only makes the region, but the entire global space industry proud." She further said, "Regional developments in the space sector in South Asia are remarkable, and no matter which nation leads - in space, all political boundaries dissolve and in space - what unites us overrides what divides us on Earth."

The Australian Space Agency also lauded the Isro's efforts and commitment to its mission. It said, "The Vikram Lander was just a few kilometres short of realising its mission to the Moon. To the team at ISRO, we applaud your efforts and the commitment to continue our journey into space."

Ironically, Pakistan's Science and Technology Minister Fawad Chaudhry shot a series of vile tweets using the hashtag "India Failed". Senator Faisal Javed Khan and DG ISPR Asif Ghafoor backed him. These men were trolled endlessly by Netizens all over the world, including Pakistani Netizens as well.

Chandrayaan-3

Chandrayaan-3 marks India's significant lunar mission, following Chandrayaan-1 and Chandrayaan-2. The mission aims to softly land on the Moon's surface, building on the previous missions. ISRO led this mission, aiming to achieve a successful landing and deploy a rover for experiments and data collection. The focus is on studying the Moon's southern pole, including its geology, minerals, and exosphere. This knowledge will deepen our understanding of the Moon's history and formation.

Chandrayaan-3 aimed to safely demonstrate a soft landing on the lunar surface, deploy a rover, and conduct scientific experiments. Advanced technologies within the Lander include laser and RF-based Altimeters, Velocimeters, and Propulsion Systems. To ensure success, rigorous tests have been conducted on Earth to simulate these conditions.

This mission showcased India's technological and scientific prowess, reflecting its commitment to space exploration. A successful Chandrayaan-3 aims to enhance India's standing in the global space community, marking another remarkable achievement in ISRO's journey.

On July 14, 2023, at 2:35 pm IST, Chandrayaan-3 was launched using the LVM3 M4 rocket from the Second Launch Pad at Satish Dhawan Space Centre in Sriharikota, Andhra Pradesh, India. After lift-off, a series of orbit-raising manoeuvres were performed using an onboard Liquid Apogee Motor and chemical thrusters to position the satellite into a Trans-lunar injection orbit. The craft circled Earth five times while incrementally increasing its orbit height, ultimately entering the Trans Lunar Injection phase on July 31, 2023.

A significant milestone was achieved on August 5, when the ISRO conducted a successful Lunar-Orbit Insertion (LOI) operation, effectively placing Chandrayaan-3 into lunar orbit. This crucial manoeuvre was executed from the ISRO Telemetry, Tracking, and Command Network (ISTRAC) based in Bengaluru. Chandrayaan-3's designated launcher, GSLV-Mk3, placed the integrated module in an Elliptic Parking Orbit (EPO) with approximately 170 x 36,500 km dimensions.

Chandrayaan-3 then completed five orbits around the Moon while gradually decreasing its altitude from its surface. On August 17, the Vikram lander, an indigenous component of Chandrayaan-3, detached from the orbiter to embark on its independent journey towards the lunar surface.

The mission entails a Lander module (LM), a Propulsion module (PM), and a Rover. Its primary objective is to develop and exhibit new technologies required for interplanetary missions. The Lander possesses the ability to gently land on a designated lunar site and release the Rover, which will conduct in-situ chemical analyses of the lunar terrain during its mobility. Both the Lander and Rover carry scientific instruments for lunar surface experiments.

The Propulsion Module's key role is to transport the Lander Module from the launch vehicle injection phase to the final lunar orbit at 100 km, after which it separates from the Lander. Additionally, the Propulsion Module hosts a scientific payload to be operated after the Lander Module separation.

Chandrayaan-3 is composed of three primary components, each playing a crucial role in the lunar mission:

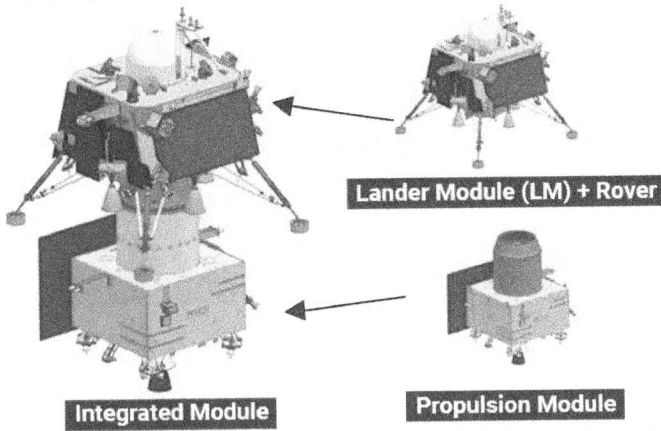

Lander Module (LM) + Rover

Integrated Module

Propulsion Module

Source: www.isro.gov.in

Orbiter Module: The propulsion module acts like a vehicle, carrying the lander and rover to a spot in space around 100 km above the Moon. This part looks like a box with a large solar panel on one side and a tall cylinder on top, known as the Intermodular Adapter Cone. The cone acts as a strong base for attaching the lander. The lander is now less heavy because some of the weight from the previous orbiter has been moved to the lander to make it tougher. Once it separates from the propulsion module, the Chandrayaan-3 lander's communication system will connect to the orbiter of Chandrayaan-2, which has been orbiting the Moon for four years. Even though the lander of the Chandrayaan-2 mission didn't land smoothly on the Moon, its orbiter part is still going around the Moon and is helping ensure Chandrayaan-3 succeeds.

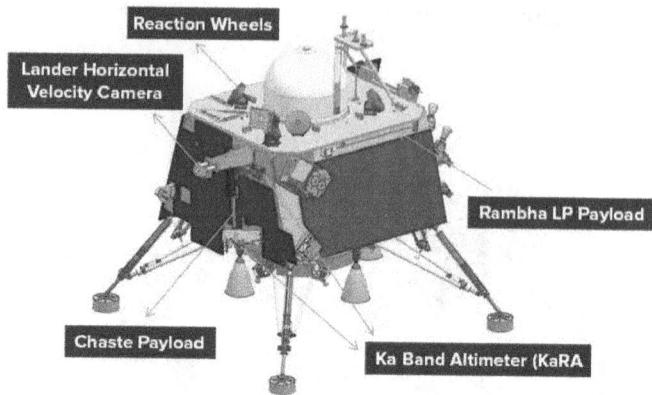

Source: www.isro.gov.in

Lander: The lander's primary task is to ensure a gentle landing on the Moon's surface. It features a box-shaped design with four landing legs and four 800 Newtons landing thrusters. The lander carries the rover and an array of scientific instruments essential for conducting on-site analyses.

Chandrayaan-3's lander is designed with notable improvements to address previous mission challenges. The lander boasts four throttle-able engines that allow for adjustments in thrust valve slew rate, unlike Chandrayaan-2's Vikram lander, which had five fixed-thrust engines. Notably, the lander can control attitude and thrust throughout its descent phases, mitigating issues faced during Chandrayaan-2's camera coasting phase. The Chandrayaan-3 lander also integrates a Laser Doppler Velocimeter (LDV) to facilitate attitude measurement in three directions. The impact legs have been reinforced compared to Chandrayaan-2, and instrumentation redundancy has increased.

Aiming for enhanced precision, the Chandrayaan-3 lander has its sights set on a targeted landing region measuring 4 km by 4 km. This area is determined based on images previously acquired by the Orbiter High-Resolution Camera (OHRC) on Chandrayaan-2. ISRO has introduced various improvements, such as strengthening structural rigidity, elevating instrument polling, boosting data frequency and transmission, and implementing multiple software and contingency systems. These enhancements are intended to bolster the lander's resilience, enabling it to endure potentially challenging landing scenarios.

The rover in Chandrayaan-3 has two important tools on board:

1. **Laser Induced Breakdown Spectroscope (LIBS):** This instrument figures out what the Moon's surface is made of, like its chemicals and minerals.

2. **Alpha Particle X-ray Spectrometer (APXS):** This tool tells us about the elements present on the Moon's surface. ISRO specifically mentions that the rover will look for elements like magnesium, aluminium, silicon, potassium, calcium, titanium, and iron.

The lander in Chandrayaan-3 carries four different tools:

1. **Radio Anatomy of Moon Bound Hypersensitive ionosphere and Atmosphere (RAMBHA):** This tool helps us understand how the gases and plasma in the local environment around the Moon change over time.

2. **Chandra's Surface Thermophysical Experiment (ChaSTE):** This tool helps scientists learn about how the Moon's surface reacts to heat.

3. **Instrument for Lunar Seismic Activity (ILSA):** ILSA measures shaking or vibrations on the Moon to understand its internal structure.

4. **Laser Retroreflector Array (LRA):** NASA provided this tool, which helps scientists measure the distance between the Earth and the Moon using lasers. This process is called laser ranging, and it involves sending a laser beam to the Moon and measuring how long it takes for the light to come back.

The lander and rover in Chandrayaan-3 are powered by solar energy. They will spend around two weeks exploring their surroundings on the Moon. However, they are not built to withstand the cold lunar nights, so their mission is limited to this time. The rover can talk only to the lander, sending messages directly to Earth. ISRO mentions that the Chandrayaan-2 orbiter could also be a backup communication link.

This successful landing was far from guaranteed, as four out of the last six lunar landing attempts in the past five years had failed. A recent instance was on August 19, when Russia's Luna-25 spacecraft suffered engine failure and crashed on the Moon. This highlighted the ongoing risks of reaching the lunar surface safely. Other unsuccessful attempts include Israel's Beresheet by SpaceIL, India's Chandrayaan-2, and Japan's Hakuto-R spacecraft by the private company Ispace. These instances emphasize the challenges of lunar exploration and the importance of careful planning and execution.

Pragyaan Rover.

The rover for the Chandrayaan-3 mission has some exciting features. It weighs about 27 kg and runs on solar power. Its size is about 0.9 meters in length, 0.75 meters in width, and 0.85 meters in height. The rover's power is 50 W, and it moves at a speed of 1 cm per second. The mission's duration is planned to be around 14 days, which is one lunar day.

This rover is built to move on six wheels and travel a distance of 500 meters on the moon's surface. It will move slowly, at 1 cm per second, while studying the environment and sending data to the Vikram lander. The lander would then transmit this data to Earth.

For navigation, the rover uses a unique system. It had two cameras at the front, called NAVCAMs, which provided a 3D view of the terrain. This camera helped the control team on Earth plan the rover's path. Scientists from IIT Kanpur helped develop parts of the system that created maps using light and designed the rover's movements.

The rover's design included a unique suspension system and six wheels. Each wheel had its own electric motor. The rover could turn by adjusting the speeds of its wheels. It was supposed to work for about 14 Earth days, which is one lunar day. This is because its electronics may not handle freezing lunar nights. It has an intelligent power system that uses solar energy and a sleep/wake-up cycle, extending its working time more than planned.

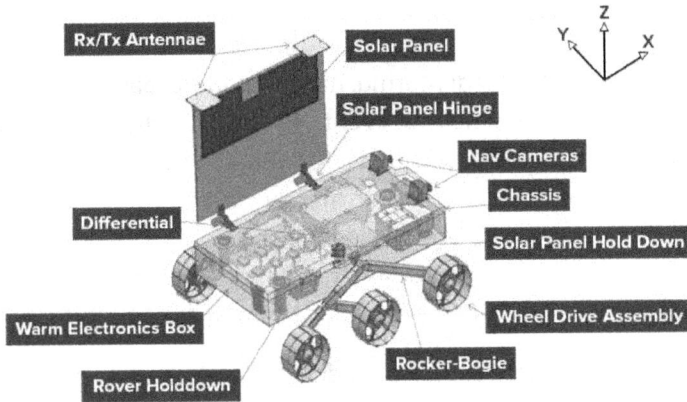

Source: www.isro.gov.in

With the Vikram lander's successful touchdown, it's now waiting for the dust to settle, quite literally. Soon, a side panel will open, and a ramp will extend for the rover named Pragyaan. This rover will slide down this ramp onto the Moon's surface. Once on the surface, Pragyaan will start its "Moonwalk" at a leisurely pace of 1 centimetre per second. It'll explore the rocky terrain and craters, collecting essential data and snapping images.

Pragyaan will communicate solely with the lander. The lander will transmit this data to the Chandrayaan-2 orbiter orbiting the Moon. From there, the information will be relayed to Earth for analysis.

This landing coincides with the start of a lunar day, which lasts 28 Earth days. During this time, the lander and rover will soak up the 14 days of sunlight to charge their batteries. When night falls, they'll temporarily stop functioning as their batteries deplete. It's still determined whether they'll reactivate during the next lunar day.

Loaded with five scientific instruments, the lander and the rover will unveil mysteries about the Moon's surface and surroundings, both above and below.

Soft landing of Chandrayan-3:

The spacecraft's Vikram lander made the soft landing at 6.04 PM (IST) on 23 August 2023, ending the disappointment over the crash-landing of the Chandrayaan-2 lander four years ago. After the Chandrayaan-3 lander made a safe touchdown on the Moon's surface on 23 August, jubilant celebrations

144

erupted throughout the nation. Messages of congratulations flooded various social media platforms.

India's President, Droupadi Murmu, hailed Chandrayaan-3's gentle landing on the Moon as a monumental event. Prime Minister Modi extended his felicitations to the ISRO scientists for the success of the Chandrayaan-3 mission. He remarked, "India isn't merely on the Moon; this is an eternal moment to treasure. We are witnessing the new flight of a new India." He also emphasized that India's lunar triumph was a global achievement, reflecting the principle of one Earth, one family, and one future.

Assistance from the US Space Agency, NASA, and the European Space Agency was crucial in tracking the Chandrayaan-3 mission during periods when India's space tracking station couldn't. Both agency heads took to social media to convey their enthusiastic congratulations.

Bill Nelson, NASA Administrator, tweeted, "Congratulations @isro on your successful Chandrayaan-3 lunar South Pole landing! And congratulations to #India on being the 4th country to successfully soft-land a spacecraft on the Moon. We're glad to be your partner on this mission!"

Josef Aschbacher, Director General of ESA, expressed admiration, "Incredible! Congratulations to @isro, #Chandrayaan_3, and to all the people of India!! What a way to demonstrate new technologies AND achieve India's first soft landing on another celestial body. Well done, I am thoroughly impressed."

The UK Space Agency joined the celebration, applauding India's achievement in becoming the fourth nation to achieve a soft moon landing.

Nepal's Prime Minister, Pushpa Kamal Dahal, took to social media to congratulate Prime Minister Narendra Modi and ISRO for the historic scientific achievement. Nepal expressed a desire to collaborate with India in sharing the benefits of space exploration.

Abdulla Shahid, Foreign Minister of Maldives, congratulated PM Modi and ISRO, highlighting the significance of this achievement for humanity.

Fawad Chaudhry, the ex-Minister of Information and Broadcasting in Pakistan, congratulated ISRO despite his earlier criticism during Chandrayaan-2's failure.

Professor Anu Ojha OBE, from the UK Space Agency, commended India for engineering and perseverance, stating that Chandrayaan-3's successful landing in the Moon's southern polar region marks a new era of space exploration.

South Africa's President Cyril Ramaphosa celebrated the achievement and its resonance within the BRICS family, expressing joy and solidarity with India.

The successful landing of Chandrayaan-3 has not only brought nationwide jubilation but has also garnered international recognition and congratulations from space agencies, governments, and leaders worldwide.

Future ISRO Missions

To enhance India's role in space exploration, ISRO has outlined a series of upcoming missions aimed at the Sun, Moon, Mars, and even crewed spaceflights. Many of these missions are set to launch soon. Let's take a closer look at ISRO's forthcoming space endeavours. Under the future missions section, the ISRO official website has listed the four missions below.

ADITYA-L1

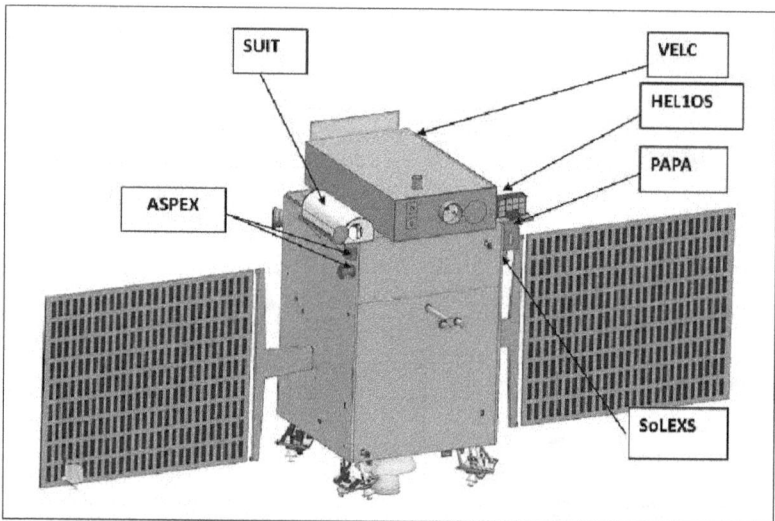

Aditya L1 marks a significant milestone as India's maiden space mission dedicated to studying the Sun. It will be positioned in outer space at the Lagrange point 1 (L1) of the Sun-Earth system, approximately 1.5 million km from Earth; the spacecraft will adopt a halo orbit. This strategic placement will allow Aditya L1 to consistently observe the Sun without interruptions caused by eclipses or occultation. Such an advantageous location will grant real-time monitoring of solar activities and their implications on space and Earth weather.

This pioneering mission encompasses seven payloads, each designed to scrutinize distinct aspects of the Sun, ranging from the photosphere and chromosphere to the outermost layer – the corona. Utilizing electromagnetic, particle, and magnetic field detectors, four payloads will directly capture solar phenomena. Concurrently, the other three payloads will undertake in-situ investigations of particles and fields at Lagrange point L1, providing valuable insights into the propagation of solar dynamics across interplanetary space.

The Aditya L1 payloads are anticipated to unravel essential insights into perplexing solar phenomena, including the enigmatic coronal heating process, dynamics of coronal mass ejections (CMEs), pre-flare and flare activities, and the behaviour of partially ionized plasma. These findings will contribute to advancing our understanding of space weather dynamics and the interaction of particles and fields emanating from the Sun. The core objectives of the Aditya L1 mission can be summarized as follows:

1. **Studying Solar Upper Atmosphere Dynamics:** A primary focus lies in comprehending the intricate dynamics of the solar upper atmosphere, encompassing the chromosphere and corona.

2. **Investigating Chromospheric and Coronal Heating:** The mission aims to shed light on the mechanisms behind coronal heating, flare initiation, and the physics of partially ionized plasma.

3. **In-Situ Particle and Plasma Environment Study:** Aditya L1 will furnish essential data for understanding particle dynamics originating from the Sun by observing the in-situ particle and plasma environment.

4. **Unlocking the Secrets of Solar Corona:** The mission seeks to decipher the mysteries surrounding the solar corona and the mechanisms responsible for its heating.

5. **Probing Plasma Diagnostics:** Aditya L1 will offer insights into the temperature, velocity, and density of the coronal and coronal loop plasma, enhancing our knowledge of these intricate processes.

6. **Unveiling the Dynamics of Coronal Mass Ejections (CMEs):** The mission will delve into the development, dynamics, and origins of CMEs, vital components of solar activity.

7. **Understanding Solar Eruptive Events:** By tracing the sequence of processes across various layers, from the chromosphere to the

extended corona, Aditya L1 aims to elucidate the genesis of solar eruptive events.

8. **Mapping Magnetic Field Topology:** The mission will offer insights into magnetic field topology and measurements within the solar corona, contributing to our comprehension of its behaviour.

9. **Exploring Space Weather Drivers:** Aditya L1 endeavours to identify the origins, composition, and dynamics of solar wind, which play a pivotal role in driving space weather phenomena.

To complete all these tasks Aditya mission payload is planned to carry the below instruments.

Equipment	Mission
Remote Sensing Payloads	
Visible Emission Line Coronagraph(VELC)	Solar Corona Imaging & Spectroscopy
Solar Ultraviolet Imaging Telescope (SUIT)	Photosphere and Chromosphere Imaging in Narrow & Broadband
Solar Low Energy X-ray Spectrometer (SoLEXS)	Solar Observation with Soft X-ray spectrometer
High Energy L1 Orbiting X-ray Spectrometer(HEL1OS)	Solar Observation with Hard X-ray spectrometer
On-Site Payloads	
Aditya Solar wind Particle Experiment(ASPEX)	Solar wind/Particle Analyzer Protons & Heavier Ions with directions
Plasma Analyser Package For Aditya (PAPA)	Solar wind/Particle Analyzer Electrons & Heavier Ions with directions
Advanced Tri-axial High Resolution Digital Magnetometers	On-Site magnetic field (Bx, By and Bz).

With its diverse payloads and ambitious objectives, the Aditya L1 mission represents India's commitment to advancing solar science and understanding the Sun's profound influence on our cosmic environment. This endeavour expands our knowledge of fundamental solar processes and contributes to the global exploration of space and its dynamic interplay with our planet.

XPoSat (X-ray Polarimeter Satellite)

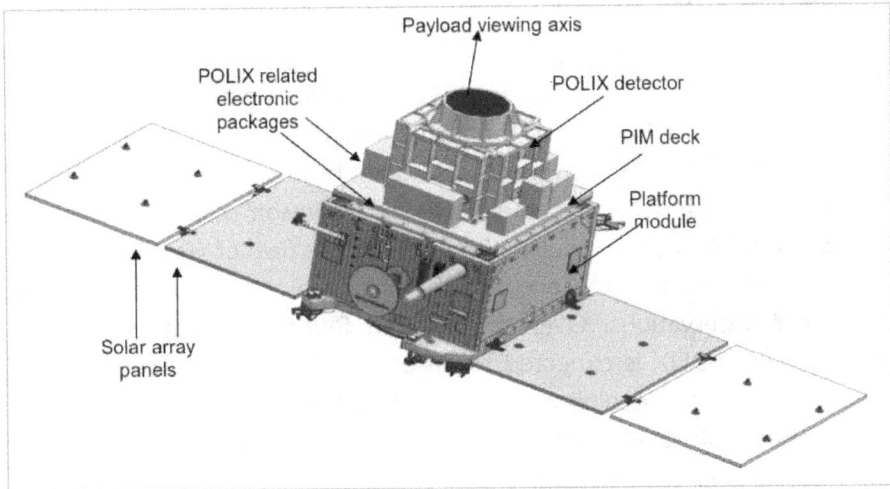

XPoSat (X-ray Polarimeter Satellite) stands as India's pioneering venture into dedicated polarimetry missions to unravel the intricate dynamics of bright astronomical X-ray sources within extreme conditions. Positioned in a low earth orbit, the spacecraft carries two vital scientific payloads that advance our understanding of celestial phenomena.

At the heart of this mission lies POLIX (Polarimeter Instrument in X-rays), the primary payload designed to operate in the medium X-ray energy range of 8-30 keV. Developed through a collaboration between the Ramam Research Institute (RRI), Bangalore, and the U R Rao Satellite Centre (URSC), POLIX is designed to measure the degree and angle of polarization in astronomical X-ray photons. This essential data helps unlock the complex emission mechanisms from diverse sources like black holes, neutron stars, active galactic nuclei, and pulsar wind nebulae. These sources exhibit intricate physical processes that challenge our comprehension. Spectroscopic and timing information gathered by space-based observatories has provided valuable insights, yet the underlying nature of such emissions remains enigmatic. This is where polarimetry measurements come into play, offering a multidimensional diagnostic tool by introducing polarization parameters. Combining polarimetric observations with spectroscopic measurements is expected to untangle the complexities and ambiguities posed by various theoretical models of astronomical emission processes. XPoSat's contribution in this direction is set to become a hallmark of Indian scientific research.

Breaking down the payloads:

- **POLIX:** POLIX, an X-ray polarimeter designed for the energy range of 8-30 keV, boasts a meticulous architecture. Comprising a collimator, scatterer, and four X-ray proportional counter detectors encircling the scatterer, POLIX's scatterer is fashioned from low atomic mass material. This characteristic triggers anisotropic Thomson scattering of incoming polarized X-rays. The collimator confines the field of view to a compact 3 degrees by 3 degrees, ensuring the presence of a single bright source in most observations. Over the planned 5-year duration of the XPoSat mission, POLIX is expected to observe around 40 diverse astronomical sources. This revolutionary payload is the maiden instrument in the medium X-ray energy range dedicated solely to polarimetry measurements.
- **XSPECT:** XSPECT, the X-ray Spectroscopy and Timing payload aboard XPoSat, takes centre stage as a facilitator of rapid timing and high-quality spectroscopic resolution in soft X-rays. With POLIX requiring prolonged observations for X-ray polarization measurements, XSPECT seizes the opportunity to monitor spectral state changes and continuum emissions long-term. This spans shifts in line flux and profile and simultaneous temporal monitoring of soft X-ray emissions within the energy range of 0.8-15 keV. An array of Swept Charge Devices (SCDs) delivers an effective area exceeding 30 cm² at six keV, accompanied by energy resolution surpassing 200 eV at six keV. Passive collimators further mitigate background noise by narrowing XSPECT's field of view. XSPECT's mission includes observing various sources, encompassing X-ray pulsars, blackhole binaries, and low-magnetic field neutron stars in LMXBs, AGNs, and Magnetars.

XPoSat, with its cutting-edge payloads and groundbreaking scientific aspirations, aligns seamlessly with India's commitment to unravelling the cosmos' enigmas. It aspires to contribute to the global scientific community by expanding our understanding of celestial dynamics and their far-reaching implications. This mission is a testament to ISRO's unwavering pursuit of excellence in space exploration and endeavouring to enrich humanity's cosmic knowledge.

NASA-ISRO SAR (NISAR) Satellite

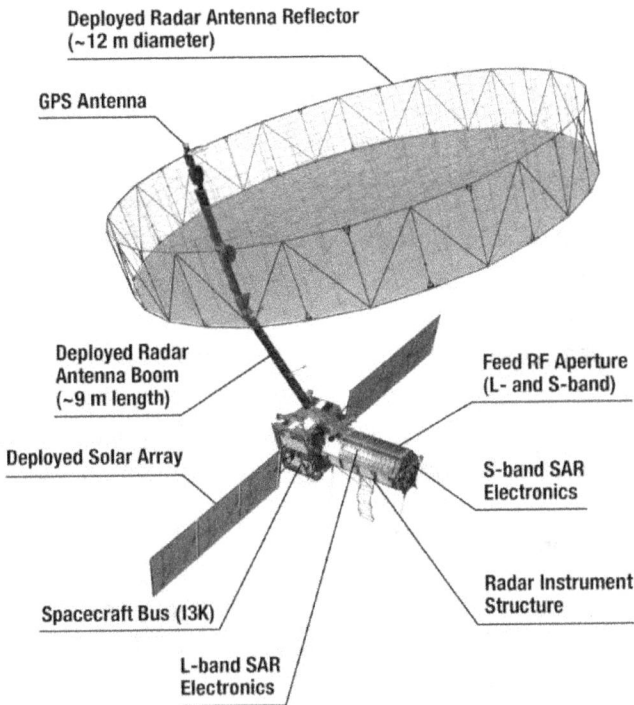

The NASA-ISRO SAR (NISAR) is a unique satellite made by NASA and ISRO. It orbits close to Earth and takes pictures of the whole world every 12 days. This helps us learn about changes in nature, like trees, ice, oceans, and even disasters like earthquakes and volcanoes. Both NASA and ISRO are working together on this big project. NASA made a radar to see things, and ISRO made another radar. These radars use a big umbrella-like antenna to talk to Earth. This partnership is essential for both NASA and ISRO. NASA is making the radar to see with L-Band, and ISRO made the S-Band radar. They both use giant umbrella antennae. NASA also brought tools like fast internet and GPS for the satellite. It's like teamwork between two space giants to help us understand Earth better.

The entire mission is divided into 3 phases. These missions are in sequence as below.

Launch Phase: The journey of the NISAR Observatory begins at Satish Dhawan Space Centre (SDSC) SHAR, Sriharikota, located on the southeastern coast of the Indian peninsula. The GSLV expendable launch vehicle provided by ISRO will carry it into space. The planned launch date is set for January 2024. This phase includes the steps from the observatory inside the launch vehicle, which is fairing on the ground, to being released into space. It concludes that the observatory is oriented towards Earth when the solar arrays are deployed, establishing two-way communication with the ground. The launch sequence is a crucial stage of the mission.

Commissioning Phase: The first 90 days after launch are dedicated to the commissioning phase, also known as in-orbit checkout (IOC). The main goal is to prepare the observatory for conducting scientific operations. Commissioning is divided into smaller steps, which include initial checkout of ISRO engineering systems and JPL engineering payload, confirming the observatory's functionality, and verifying the instruments' performance. This is a gradual process to ensure that the observatory can operate effectively, from deploying necessary parts and checking systems to testing the radars independently and jointly.

Science Operations Phase: After commissioning, the science operations phase commences and lasts for three years. During this time, the observatory collects data necessary to achieve the L1 science objectives. Regular manoeuvres are conducted carefully to maintain its science orbit to avoid disrupting science observations. The initial months focus on thorough calibration and validation activities, with yearly updates lasting a month each. The plan for observing both L- and S-band instruments and engineering tasks is established before launch through coordination between JPL and ISRO. This is known as the reference mission. The specific science observations within this reference mission are part of the reference observation plan (ROP). The schedule for science observations is determined using L- and S-band target maps, radar mode tables, and considerations about the spacecraft and ground-station capabilities. JPL's mission planning team manages this schedule, aiming to follow the reference mission's plan closely, accounting for minor timing adjustments based on the actual orbit.

SPADEX-Third quarter of 2024.

SPADEX, known as the Space Docking Experiment, is an ISRO project focused on advancing technologies connected to tasks like orbital rendezvous, docking, formation flying, and proximity operations in space. These technologies have wide-ranging applications, including human spaceflight, satellite servicing in space, and other close interactions between spacecraft.

The Space Docking Experiment is planned to take off from the Satish Dhawan Space Centre using a Polar Satellite Launch Vehicle. In this mission, two satellites of the IMS class (each weighing around 200 kg) will be used – one called Chaser and the other named Target. Both satellites will be launched together on a single mission but placed into slightly different orbits. The primary objectives of this mission are to achieve autonomous rendezvous and docking of the two spacecraft and demonstrate the ability to control one spacecraft using the attitude control system of the other. At the same time, they are docked, showcase formation flying capabilities, and test remote robotic arm operations in space.

Initial steps for the Space Docking Experiment were taken in 2016, and the Government of India gave it the green light. In 2017, an initial fund of Rs 10 crore was approved for the project. Subsequently, in June 2019, ISRO sought proposals to study technologies related to remote robotic arm operations, rendezvous, and docking on its PSLV fourth stage (PS4) orbital platform. Although the ISRO website did not specify any dates for the mission, various sources claim that the mission will start in the middle of 2024.

The below projects are not listed on the official ISRO website, but various sources have confirmed that work on the following programs is on the way.

Gaganyaan

The Gaganyaan project aims to demonstrate India's capability for human spaceflight. It involves launching a crew of 3 members into an orbit 400 km above Earth for a 3-day mission, safely returning them by landing in the Indian Ocean.

To achieve this, a strategic approach combines in-house expertise, Indian industry experience, academic and research institution capabilities, and advanced technologies from international agencies. The project requires developing critical technologies, including a human-rated launch vehicle for safe crew transport, a Life Support System to create a habitable environment in space, provisions for crew emergency escape, and comprehensive crew management strategies for training, recovery, and rehabilitation. Before the human spaceflight, several precursor missions are planned to demonstrate Technology Preparedness Levels. These include the Integrated AirDrop Test (IADT), Pad Abort Test (PAT), and Test Vehicle (TV) flights. These unmanned missions will validate system safety and reliability before the manned mission.

Gaganyaan, India's inaugural human space mission, is scheduled for launch in 2024. The unmanned 'G1 mission' is set to launch in the last quarter of 2023, followed by the second unmanned 'G2 mission' in the second quarter of 2024. The final phase, the human space flight 'H1 mission,' is planned for the last quarter of 2024. The trusted LVM3 rocket, ISRO's heavy-lift launcher, has been chosen as the launch vehicle for the Gaganyaan mission.

Crew Module: The Gaganyaan crew module is a self-sufficient spacecraft weighing 5.3 tonnes. It's designed to carry a team of three astronauts into orbit and ensure their safe return to Earth, with a mission lasting up to 3 days. The crew module has two parachutes for backup, though one parachute is sufficient for a secure splashdown in the Indian Ocean. These parachutes slow down the module during its ocean landing. The spacecraft has life support systems and controls to maintain a suitable environment. It also features an emergency mission abort capability and a Crew Escape System (CES) that can be activated during the initial or second rocket stage.

Service Module: The service module, weighing 2.9 tonnes, is powered by liquid propellant engines. When combined with the crew module, it forms an 8.2-tonne orbital module. The Service Module Propulsion System (SMPS) employs liquid propellants MON-3 and Monomethylhydrazine for orbit-raising manoeuvres. This enables Gaganyaan to reach a low earth orbit (LEO) of 400 km and remain connected for deorbiting until reentry into the Earth's atmosphere. The SMPS is outfitted with five main engines derived from ISRO's liquid apogee motor, producing a 440 N thrust and sixteen 100 N reaction control system (RCS) thrusters for precise control.

Shukrayaan

Building on the accomplishments of the Chandrayaan and Mars Orbiter Mission, ISRO has been evaluating the possibility of future interplanetary journeys to planets like Venus, our closest celestial neighbour. The proposed Venus Orbiter Mission, Shukrayaan in India, aims to send an orbiter to Venus to explore its surface and atmosphere.

The mission idea was introduced during a space conference in Tirupati in 2012. The Indian government allocated a 23% budget increase to the Department of Space in the 2017-18 budget, focusing on space science. The budget specifically mentioned allocations for Mars Orbiter Mission II and a Venus mission, thereby authorizing the initiation of preliminary studies. Collaboration was initiated with the Japan Aerospace Exploration Agency (JAXA) from 2016 to 2017 to analyze Venus's atmosphere using data from Akatsuki's radio occultation experiment. An 'Announcement of Opportunity' was made by ISRO on April 19, 2017, inviting science payload proposals from Indian academia based on the mission's general criteria. Subsequently, on November 6, 2018, another opportunity was extended to the international scientific community to propose payloads. The available payload capacity was revised to 100 kg from the initial 175 kg.

In 2018, discussions were held between the space agencies of India (ISRO) and France (CNES) about possible collaboration on this mission, focusing on autonomous navigation and aerobraking technologies. Jacques Blamont, a French astrophysicist experienced in the Vega program, suggested using inflated balloons to study the Venusian atmosphere. ISRO expressed interest in this proposal and considered deploying instrumented balloons from an orbiter, mirroring the approach used during the Vega missions. By late 2018, the Venus mission was in the configuration study phase, although full approval from the Indian government had not been sought. Plans were discussed to include a drone-like probe in the mission, according to Somak Raychaudhury, director of IUCAA. An update shared with NASA's Decadal Planetary Science Committee by ISRO scientist T Maria Antonita indicated a projected launch in December 2024, with a backup window in 2026. By November 2020, ISRO had identified 20 international proposals for collaboration involving countries such as Russia, France, Sweden, and Germany. The Swedish Institute of Space Physics partnered with ISRO for the Venus Orbiter Mission.

In May 2022, ISRO chairman S. Somanath confirmed the mission's planned launch in December 2024, with an alternative launch window of 2031.

Conclusion

As we conclude our journey through the remarkable story of the Indian Space Research Organisation (ISRO), we are reminded that the pursuit of knowledge knows no bounds. ISRO's unwavering dedication to pushing the boundaries of space exploration has not only propelled India to the forefront of scientific advancement but has also ignited a spark of inspiration that reaches far beyond Indian borders.

The tale of ISRO's achievements is a testament to human ingenuity, collaboration, and an insatiable curiosity to explore the cosmos. From launching satellites for communication, navigation, and weather forecasting to embarking on interplanetary missions that have expanded our understanding of neighbouring planets, ISRO's endeavours have left an indelible mark on both scientific progress and societal development.

But as we stand at this juncture, our journey is far from over. The stars beckon us with mysteries yet to be unravelled, and the universe holds secrets that await our discovery. ISRO's story is not just a chronicle of the past but a launchpad for the future. The organization's ambitions remain as bold as ever, with upcoming missions that will further our understanding of space, technology, and our place in the cosmos.

I hope this book has been informative and fun to read. This book will also give you all the information about ISRO's history and activities up to the Chandrayaan-3 mission. ISRO's journey will inspire many young minds to take STEM (Science, Technology, Engineering, and Mathematics) as a career choice.

In this digital age, staying connected has always been challenging. ISRO's website, social media channels, and various science outreach platforms offer windows into the world of space exploration. By following these channels, attending lectures, and participating in space-related events, we can foster a sense of camaraderie with the passionate community of scientists, engineers, and enthusiasts who share our fascination with the cosmos.

But our role doesn't stop at observation; it extends to action. Support for science education and engagement initiatives can make a profound difference in nurturing the next generation of scientific minds. By mentoring young scientists, participating in science communication, and advocating for continued investment in space research, we become integral partners in shaping the trajectory of exploration.

I have ensured that the information in this book is accurate to the best of my knowledge and research. If you find any errors in this book, please feel free to share them with me at https://cyberauthor.tech/ so I can make the necessary corrections in future editions. I also encourage readers to share their feedback with me. This will help me improve my writing skills so I can bring more exciting topics for you to read in the future.

As we close this chapter, let us remember that the legacy of ISRO is not confined to the pages of history—it is a living, breathing testament to human potential. Together, we can celebrate the achievements of ISRO while fueling the fire of curiosity that drives us to reach for the stars. The universe awaits, and our journey has only just begun.

In the words of Dr. A.P.J. Abdul Kalam, a luminary who was both a space scientist and the President of India, "We are all born with a divine fire in us. Our efforts should be to give wings to this fire and fill the world with the glow of its goodness." Let us embrace this fire and collectively illuminate the path to a future among the stars.

Keep looking up, stay curious, and let the spirit of exploration guide our way.

Thank you for reading this book. Jai Hind.

www.ingramcontent.com/pod-product-compliance
Lightning Source LLC
Chambersburg PA
CBHW030331220326
41518CB00048B/2230